REVISE OCR AS/A LEVEL
Chemistry

REVISION WORKBOOK

Series Consultant: Harry Smith

Author: Mark Grinsell

Our revision resources are the smart choice for those revising for OCR AS/A Level Chemistry. This book will help you to:

- **Organise** your revision with the one-topic-per-page format
- **Prepare** for your AS/A Level exam with a book full of exam-style practice questions
- **Simplify** your revision by writing straight into the book just as you would in an exam
- **Track** your progress with at-a-glance check boxes
- **Improve** your understanding, and exam technique, with guided questions to build confidence, and hints to support key revision points.

Revision is more than just this Workbook!

Make sure that you have practised every topic covered in this book, with the accompanying OCR AS/A Level Chemistry Revision Guide. It gives you:

- A 1-to-1 page match with this Workbook
- Explanations of key concepts delivered in short memorable chunks
- Key hints and tips to reinforce your learning
- Worked examples showing you how to lay out your answers
- Exam-style practice questions with answers.

For the full range of Pearson revision titles across GCSE, BTEC and AS/A Level visit:
www.pearsonschools.co.uk/revise

ALWAYS LEARNING

PEARSON

Published by Pearson Education Limited, 80 Strand, London, WC2R 0RL.

www.pearsonschoolsandfecolleges.co.uk

Copies of official specifications for all OCR qualifications may be found on the OCR website:
www.ocr.org.uk

Text and illustrations © Pearson Education Limited 2016
Copyedited by Ros Woodward and Hilary Herrick
Typeset by Kamae Design
Produced by Out of House Publishing
Illustrated by Tech-Set Ltd, Gateshead
Cover illustration © Miriam Sturdee

The rights of Mark Grinsell to be identified as author of this work has been asserted by him in accordance with the Copyright, Designs and Patents Act 1988.

First published 2016

19 18 17 16
10 9 8 7 6 5 4 3

British Library Cataloguing in Publication Data
A catalogue record for this book is available from the British Library

ISBN 978 1 447 98432 0

Printed in the UK by Bell and Bain Ltd, Glasgow

Contents

1-to-1 page match with the OCR Chemistry Revision Guide ISBN 9781447984375

Atomic structure and isotopes

1   There are three naturally occurring isotopes of hydrogen, ^1H, ^2H and ^3H. The percentage composition of atoms in naturally occurring hydrogen is given.

Isotope	%
^1H	99.99
^2H	0.01
^3H	< 0.01

(a) Give the meaning of **isotopes**.

.....Isotopes are atoms of the same element with.
...different numbers of neutrons and different.........
.masses.'.. **(2 marks)**

(b) Give **two** similarities in the sub-atomic particles in a ^2H atom and in a ^3H atom.

....

.... **(2 marks)**

(c) What is unusual, in terms of sub-atomic particles, about atoms of ^1H and ^2He?

How many of each sub-atomic particle are found in ^1H and ^2He?

.... **(1 mark)**

(d) The relative atomic mass is the mean relative mass of an atom.
The relative atomic mass of hydrogen, to three decimal places, is most likely to be:

☐ A   1.000

☐ B   1.008

☐ C   1.500

☐ D   2.000   **(1 mark)**

Maths skills The mean is the average of the masses, taking into account the % composition.

(e) (i) Explain why the chemical reactions of pure samples of hydrogen gas, $^1H_2$, $^2H_2$ and $^3H_2$, would be the same.

....

....

.... **(2 marks)**

> Guided

(ii) Suggest whether the rate of reaction would be different if the reactions of each of these three isotopes of hydrogen were monitored.

The molecules $^3H_2$ are heavier so would move ....

....

....

.... **(2 marks)**

Relative masses

1 (a) What is the meaning of **relative isotopic mass**?

....

....

.... **(2 marks)**

(b) Chlorine exists as two isotopes, ^{35}Cl and ^{37}Cl.

(i) State the number of protons, number of neutrons and number of electrons found in one atom of each of these isotopes.

^{35}Cl ....

^{37}Cl .... **(4 marks)**

> **Guided**

(ii) The relative atomic mass of chlorine is 35.5. State the meaning of **relative atomic mass.**

This is the weighted .... **(1 mark)**

(iii) Explain why the relative atomic mass of chlorine is closer to 35 than 37.

> How is the mean calculated?

....

....

.... **(1 mark)**

2 The relative formula mass of calcium oxide is 56 and the relative molecular mass of carbon dioxide is 44. When calcium carbonate decomposes the equation is $CaCO_3 \rightarrow CaO + CO_2$.

A   A molecule of calcium oxide is $4\frac{2}{3}$ times heavier than a carbon atom.

B   A molecule of carbon dioxide is 44 times heavier than a carbon atom.

C   100 g of calcium carbonate would make 56 g calcium oxide and 44 g carbon dioxide.

Which of the statement(s) are correct?

☐ 1   only A

☐ 2   only B

☐ 3   only C

☐ 4   A, B and C   **(1 mark)**

> Use your data sheet to find relative atomic masses.

> **Maths skills**

3 Calculate the relative molecular or formula masses of

(a) Cl_2 .... **(1 mark)**

(b) $Cu(NO_3)_2$ .... **(1 mark)**

(c) $CoCl_2.6H_2O$ .... **(1 mark)**

Using mass spectroscopy

1   The mass spectrum of bromine, Br_2, is shown.

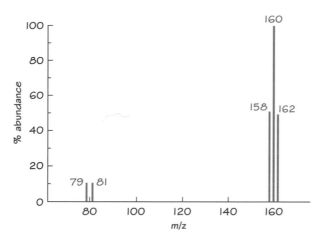

> Remember that neutral particles are not detected.

(a) The two peaks at the left are caused by bromine atoms.

   (i)   Give the formula of the species that causes the peak at 79.

.... **(1 mark)**

   (ii) Explain why there are two peaks, one at 79 and one at 81.

.... **(1 mark)**

   (iii) The peaks at 79 and 81 are of equal height.

      Explain why.

.... **(1 mark)**

   (iv) Use the information in (a)(iii) to state the relative atomic mass of bromine.

.... **(1 mark)**

(b) The three peaks on the right are caused by bromine molecules.

   (i)   Give the formula of the species that causes the peak at 158.

.... **(1 mark)**

   (ii) Give the formula of the species that causes the peak at 160.

.... **(1 mark)**

   (iii) Give the formula of the species that causes the peak at 162.

.... **(1 mark)**

Maths skills

2   Neon has three isotopes, ^{20}Ne (90.48%), ^{21}Ne (0.27%) and ^{22}Ne (9.25%).

Calculate the relative atomic mass of neon, giving your answer to two decimal places.

....

.... **(2 marks)**

Writing formulae and equations

1   When silver nitrate solution is added to sodium carbonate solution, a precipitate of silver carbonate is formed.

(a) Give the name of the second product.

.... **(1 mark)**

(b) Write down the formulae of each of the substances.

silver nitrate ....

sodium carbonate ....

silver carbonate ....

second product .... **(2 marks)**

(c) Write the balanced equation for the reaction.

.... **(1 mark)**

2   A compound of a metal, X, and a non-metal, Y, has the formula X_3Y_2.
X and Y could be

	X	Y
☐ A	Na	O
☐ B	Ca	S
☐ C	Al	N
☐ D	Mg	P

(1 mark)

3   Write balanced equations for the following reactions.

(a) Ethanol, C_2H_5OH, fully burns in air.

.... **(3 marks)**

> The carbon and the hydrogen are both fully oxidised.

(b) Carbon dioxide reacts with calcium hydroxide solution to form a precipitate of calcium carbonate and one other compound.

.... **(3 marks)**

> **Guided**

(c) Bromine displaces iodine from potassium iodide solution.

Br_2 + ... **(3 marks)**

(d) A Group 2 metal, M, has an oxide which reacts with hydrochloric acid to make the chloride of the metal and water.

.... **(3 marks)**

Amount of substance – the mole

1 (a) Give the definition of a mole.

....

....

....   **(2 marks)**

(b) The Avogadro constant is $6 \times 10^{23}\,mol^{-1}$.

Which of the following statements is true?

☐ A   There are 6×10^{23} atoms in one mole of carbon dioxide, CO_2.

☐ B   One carbon-12 atom has a mass $\dfrac{1}{(6 \times 10^{23})}$ g.

☐ C   There are 6×10^{23} molecules in one mole of water, H_2O.

☐ D 6×10^{23} atoms of any substance have the same mass.   **(1 mark)**

[Maths skills]

2 (a) Calculate the mass of

(i)   1 mole of ethanol, C_2H_5OH ....   **(1 mark)**

(ii) 0.05 mol of sodium, Na ....   **(1 mark)**

(b) Calculate the amount, in mol, of

(i)   1.55 g phosphorus atoms, P ....   **(1 mark)**

(ii) 2.00 g oxygen, O_2 ....   **(1 mark)**

>Guided>

(c) (i) 1.55 g phosphorus reacts with exactly 2.00 g oxygen gas. Using your answers to (b), calculate the formula of the phosphorus oxide formed.

> Work out the ratio using oxygen **atoms**, as oxygen atoms appear in formulae (not molecules).

Ratio of moles P : moles O = ...

....

....

....   **(2 marks)**

(ii) Write the equation for the reaction in (i).

....   **(1 mark)**

3 0.025 mol of a mole of a diatomic element has a mass of 4.00 g.
Identify the element.

....

....

....

....   **(2 marks)**

Calculating reacting masses and gas volumes

One mole of any gas has a volume of $24\,dm^3$ at room temperature and pressure.
The Avogadro constant is $6 \times 10^{23}\,mol^{-1}$.

Maths skills

1 (a) Which of the following has the largest amount of substance (number of moles)?

☐ A $10\,g$ of sodium hydroxide, NaOH

☐ B $11\,g$ of carbon dioxide, CO_2

☐ C $12\,000\,cm^3$ of hydrogen, H_2

☐ D 1.5×10^{23} atoms of argon, Ar   **(1 mark)**

(b) Which of the following has the largest mass?

☐ A $0.1\,mol$ of calcium carbonate, $CaCO_3$

☐ B $1\,mol$ of carbon-13

☐ C $8\,dm^3$ of oxygen, O_2

☐ D 3×10^{23} molecules of carbon dioxide, CO_2 **(1 mark)**

2 Hydrazine azide, N_5H_5, has been investigated as a rocket fuel. It decomposes as follows:

$12N_5H_5 \rightarrow 3N_2H_4 + 16NH_3 + \ldots N_2$

(a) Complete the equation.   **(1 mark)**

Maths skills

(b) Calculate the amount of substance, in mol, in $100\,g$ hydrazine azide.

....

.... amount = ... mol   **(2 marks)**

Maths skills

Guided

(c) Calculate the maximum volume of ammonia gas formed by the decomposition of $100\,g$ hydrazine azide.

Ratio of ammonia : hydrazine azide = ...

Amount in mol of ammonia = ...

Volume of ammonia = ...   **(3 marks)**

(d) Suggest two reasons why hydrazine azide may be useful as a rocket fuel.

....

The rocket must be propelled rapidly into the atmosphere.

....

....   **(2 marks)**

Types of formulae

1 The diagram shows the representation of part of a crystal of an ionic compound.

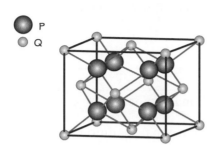

● P
○ Q

(a) What is the empirical formula of this compound?

☐ A PQ

☐ B PQ_2

☐ C P_2Q

☐ D PQ_3 **(1 mark)**

> Count the number of P and the number of Q ions shown.

(b) What is meant by **empirical formula**?

....

....

.... **(2 marks)**

Maths skills

2 A compound of carbon, hydrogen and oxygen contains 40% carbon and 6.67% hydrogen by mass.

(a) Calculate the empirical formula of the substance.

....

....

.... **(4 marks)**

(b) The relative molecular mass of the compound is 180.
    Calculate the molecular formula of the substance.

....

.... **(2 marks)**

(c) State the molecular and the empirical formulae of the molecule shown.

```
     CH₃  CH₃
      |    |
 H —— C —— C —— H
      |    |
     CH₃  CH₃
```

molecular formula .... empirical formula ....   **(2 marks)**

Calculations involving solutions

Maths skills

1 Some solid sodium carbonate is transferred to a watch glass and the mass measured. The sodium carbonate is then added to a beaker, and the watch glass re-weighed. The solid is then dissolved and made up to 250 cm^3 solution.

The readings are

mass of watch glass and sodium carbonate/ g	20.801
mass of watch glass after adding sodium carbonate to beaker/ g	17.756

(a) Calculate the concentration of the sodium carbonate solution.

....

....

.... concentration = ... mol dm^{-3} **(3 marks)**

(b) A different solution of sodium carbonate of concentration 0.200 mol dm^{-3} is used to react with 50 cm^3 of 0.300 mol dm^{-3} hydrochloric acid.

> Write the equation for the reaction first.

Calculate the volume of sodium carbonate solution required.

....

....

....

.... volume = ... cm^3 **(4 marks)**

Maths skills

2 If 4 g of sodium hydroxide are dissolved to make 100 cm^3 solution, the concentration of the solution formed in, mol dm^{-3}, is

☐ A   0.0001

☐ B   0.04

☐ C   1

☐ D   25   **(1 mark)**

Maths skills

Guided

3 100 cm^3 of hydrogen chloride gas is dissolved in water to make 500 cm^3 solution. Calculate the concentration of hydrochloric acid formed.

One mole of any gas has a volume of 24 dm^3 at room temperature and pressure.

Amount of HCl gas in mol = ...

Volume of solution in dm$_3$ = ...

Concentration = ...   **(3 marks)**

Formulae of hydrated salts

1 Epsom salt is the hydrated form of magnesium sulfate, $MgSO_4$.

(a) What is meant by hydrated?

....

.... **(1 mark)**

(b) A student carries out an experiment to determine the formula of the hydrated salt. The method was given in the student's report.

> 1. Measure the mass of a clean, empty crucible.
> 2. Add some of the hydrated magnesium sulfate and re-measure the mass of the crucible.
> 3. Heat the crucible for five minutes, allow to cool, and measure the mass.
> 4. Heat for five more minutes, allow to cool, and measure the mass.
> 5. Heat for a third time, allow to cool, and measure the mass.

Results

	mass/ g
empty crucible	11.20
crucible containing hydrated magnesium sulfate	32.24
crucible and contents after five minutes heating	23.58
crucible and contents after 10 minutes heating	23.26
crucible and contents after 15 minutes heating	23.26

(i) The crucible is washed before step 1 to ensure that the crucible is clean. Explain why it is important to dry the crucible.

.... **(1 mark)**

(ii) Explain why the heating is repeated in steps 4 and 5.

.... **(1 mark)**

(iii) Calculate the mass of anhydrous magnesium sulfate.

.... **(1 mark)**

(iv) Calculate the mass of water released.

.... **(1 mark)**

(v) Use your answers to (iii) and (iv) to find the formula of the hydrated salt.

....

....

.... **(3 marks)**

Percentage yield and atom economy

1 Consider the reaction to make ammonium chloride:

$$NH_3(g) + HCl(g) \rightarrow NH_4Cl(s)$$

Under conditions where 1 mol gas has a volume $24.0\,dm^3$, $100\,dm^3$ hydrogen chloride gas was reacted with an excess of ammonia. $201.6\,g$ ammonium chloride is formed.

(a) Calculate the atom economy of this reaction.

.... atom economy = ... %   **(1 mark)**

Guided   (b) Calculate the % yield of this reaction.

Amount HCl = ... = ... mol

Maximum amount NH_4Cl = ... mol

Maximum mass NH_4Cl = ... = ... g

Percentage yield = ... = ... %

(4 marks)

2 In the laboratory synthesis of aspirin in a flask, the following reaction is used.

aspirin
ethanoic acid

The two hexagonal rings in the equation have the formula C_6H_4.

(a) Calculate the atom economy of this reaction.

....

....

.... atom economy = ... %   **(3 marks)**

> You only need to calculate M_r for the two products.

(b) A description of the purification of the product is given.

> Ethanoic acid is very soluble in water but aspirin is not, so after the reaction, the mixture in the flask is shaken with cold water and filtered. Purified aspirin can be obtained through recrystallisation of the solid with hot ethanol.

An **incorrect** reason why the yield of aspirin in the synthesis is less than 100% is that

☐ A some aspirin is left in the flask when the mixture is filtered

☐ B some aspirin does not pass through the filter paper into the filtrate

☐ C the reaction forming aspirin may be incomplete

☐ D in recrystallisation, some aspirin will stay dissolved in the ethanol.   **(1 mark)**

Neutralisation reactions

1   Nitric acid is described as a strong acid, and ethanoic acid as a weak acid.

   (a)   Explain the difference between a strong and a weak acid.

     ....

     ....

     ....   **(2 marks)**

   (b)   The ionic equation for the reaction of nitric acid with potassium hydroxide is

     ☐   A $HNO_3 + OH^- \rightarrow NO_3^- + H_2O$

     ☐   B $H^+ + KOH \rightarrow K^+ + H_2O$

     ☐   C $H^+ + OH^- \rightarrow H_2O$

     ☐   D $H^+ + NO_3^- + K^+ + OH^- \rightarrow K^+ + NO_3^- + H_2O$ **(1 mark)**

   (c)   (i)   Write the balanced equation for the reaction of ethanoic acid with sodium carbonate.

       ....   **(3 marks)**

     (ii)   Describe the observations that you would make if carrying out the reaction in (c)(i) in the laboratory.

> Sodium ethanoate is soluble in water.

       ....

       ....   **(2 marks)**

Practical skills

2   You are given unlabelled bottles of three colourless solutions, each of a high concentration. One is sodium carbonate solution, one is potassium hydroxide solution and one is dilute sulfuric acid.

   (a)   Using test tubes and a thermometer only, describe how you would identify the three solutions.

     ....

     ....

     ....

     ....

     ....

     ....   **(4 marks)**

   (b)   Explain a safety precaution you should use when carrying out the experiment.

     ....

     ....   **(1 mark)**

Acid–base titrations

1   A standard solution of sodium hydrogensulfate, $NaHSO_4$, is made by the procedure indicated below.

> 1. A container is weighed, and then approximately 3 g of anhydrous sodium hydrogensulfate is added to the container.
> 2. The solid is tipped into a small beaker, and the container is reweighed.
> 3. The solid is dissolved in distilled water.
> 4. The solution is poured, using a funnel, into a 250 cm³ volumetric flask.
> 5. The volumetric flask is made up to the mark with distilled water.

(a) Explain why the container is reweighed in step 2, rather than using the mass of the container measured in step 1.

....

.... **(1 mark)**

(b) Give an improvement to the method described in step 4.

....

.... **(1 mark)**

(c) Give an improvement to the method described in step 5.

....

.... **(1 mark)**

Maths skills

(d) Calculate the concentration of the solution using the results below.

Mass of container and contents/ g	23.105
Mass of emptied container/ g	20.010

....

....

.... concentration = ... mol dm⁻³   **(3 marks)**

(e) In a titration, sodium hydrogensulfate is pipetted into a flask and sodium hydroxide solution is added from a burette until the end point.

(i)   Write the balanced equation for the reaction.

.... **(2 marks)**

(ii) Give the colour change at the end point if the indicator used is phenolphthalein.

.... **(1 mark)**

Calculations based on titration data

1 In a titration, 25.0 cm³ of sulfuric acid is pipetted into a flask. An indicator is added and the acid is titrated with 0.110 mol dm⁻³ potassium hydroxide solution. The volumes of potassium hydroxide are given.

Experiment	1	2	3	4
Final volume/ cm³	18.05	21.20	23.45	25.65
Initial volume/ cm³	0.00	2.20	5.35	6.90
Titre/ cm³				

(a) (i) Complete the table to give the titre values. **(2 marks)**

(ii) Using suitable values, calculate the mean titre.

.... mean titre = ... cm³   **(2 marks)**

(iii) Calculate the amount of potassium hydroxide in the mean titre.

.... amount KOH = ... mol   **(2 marks)**

(b) (i) Write the balanced equation for the reaction.

....   **(2 marks)**

>**Guided**

(ii) Calculate the amount of sulfuric acid that reacts with the amount of KOH in (a) (iii).

A 1:2 ratio of .... amount H₂SO₄ = ... mol   **(2 marks)**

(c) Calculate the concentration of the sulfuric acid.

.... [H₂SO₄] = ... mol dm⁻³   **(2 marks)**

(d) In the titration, the flask used is rinsed with water between each titre.

> Consider whether any acid or alkali is being added.

Which statement is correct about the calculated value of the concentration of sulfuric acid?

☐ A If the water is pure, it will have no effect on the result.

☐ B If the water is pure, the calculated concentration of the acid will be too high.

☐ C If the water is pure, the calculated concentration of the acid will be too low.

☐ D If the water is acidic, the calculated concentration of the acid will be too low.   **(1 mark)**

(e) Explain why the concentration of potassium hydroxide used in the titration calculation must be found just before use, and would be incorrect if the solution is left for several days.

> The KOH will react with acidic gases.

....   **(1 mark)**

Oxidation numbers

1 State the oxidation number of chlorine in each of the following substances.

KCl ....

$KClO_3$ ....

ClF_3 ....

Cl_2 ....

(4 marks)

2 A substance in which the oxidation number of oxygen is –1:

☐ A is K_2O

☐ B is MgO

☐ C is H_2O_2

☐ D does not exist because the oxidation number is always –2. **(1 mark)**

3 Peroxodisulfate ions, $S_2O_8^{2-}$, react with iron(II) ions, Fe^{2+}, to make sulfate ions and iron(III) ions.

(a) (i) Write the half-equation for the oxidation of the Fe^{2+} ions.

.... **(1 mark)**

(ii) Write the half-equation for the reduction of the $S_2O_8^{2}$ ions.

.... **(2 marks)**

> Balance the atoms first and then add the number of electrons to balance the charges.

(b) (i) Give the change in oxidation number of iron in the equation in (a)(i)

from .... to .... **(1 mark)**

(ii) Give the oxidation number of sulfur in SO_4^{2-}, and hence the systematic name of the ion.

.... **(2 marks)**

(c) Write the overall ionic equation for the reaction.

.... **(1 mark)**

> **Guided**

4 Explain why in the compound ClF chlorine is given a positive oxidation number, +1.

Fluorine's electronegativity is ....

....

....

.... **(2 marks)**

14

Examples of redox reactions

1 Write the formulae of the following compounds

   (a) iron(III) oxide ....   **(1 mark)**

> The oxidation number can be used as the charge on the iron.

   (b) iron(II) oxide ....   **(1 mark)**

   (c) The three irons in Fe_3O_4 have oxidation numbers of

      ☐   A   +2, +2, +2

      ☐   B   +2, +2, +4

      ☐   C   +2, +3, +3

      ☐   D   +3, +3, +3   **(1 mark)**

2 Potassium dichromate, $K_2Cr_2O_7$, in acid solution is used to oxidise primary alcohols. In the reaction, chromium(III) ions, Cr^{3+}, are formed.

   (a) Give the IUPAC name of $K_2Cr_2O_7$, including the relevant oxidation number.

   ....   **(1 mark)**

   (b) Write the half-equation for the reduction of the dichromate ions.

   ....   **(3 marks)**

   (c) The ion VO_2^+ can be reduced in sequence to VO^{2+} then $[V(H_2O)_6]^{3+}$ and finally $[V(H_2O)_6]^{2+}$ by zinc in acid solution.

      Give the oxidation state of **vanadium** in each of the species.

      VO_2^+ .... VO^{2+} ....

      $V(H_2O)_6]^{3+}$ .... $[V(H_2O)_6]^{2+}$ ....   **(4 marks)**

> Guided

3 When copper(I) oxide is added to dilute sulfuric acid, a red-brown precipitate and a blue solution are formed in a disproportionation reaction.

   (a) What is **disproportionation**?

   The same element is ....

   ....   **(1 mark)**

   (b) Explain, in terms of oxidation and reduction, what reaction is occurring.

   ....

   ....

   ....   **(3 marks)**

Exam skills 1

Maths skills

1 (a) The substance **R** contains 34.33% sodium, 17.91% carbon and oxygen.

(i) Calculate the empirical formula of **R**.

....

....

....

.... **(3 marks)**

(ii) The relative formula mass of **R** is 134.

Find the formula of **R**.

.... **(1 mark)**

(b) **R** reacts with $KMnO_4$ in acid solution.

> The H^+ ions form water in this reaction.

(i) Construct the half equation for the reduction of manganate ions, MnO_4^-, in acid solution, forming Mn^{2+} ions.

....

.... **(2 marks)**

(ii) Give the oxidation state of manganese in the two ions

MnO_4^- .... **(1 mark)**

Mn^{2+} .... **(1 mark)**

(c) **R** decomposes to form sodium carbonate, Na_2CO_3 and a toxic gas only.

> Use your formula from (a)(ii).

(i) Write the equation for the reaction.

.... **(2 marks)**

(ii) Calculate the atom economy for the formation of sodium carbonate in this reaction.

....

....

.... **(3 marks)**

(iii) If **R** is strongly heated to carry out this decomposition, the percentage yield of sodium carbonate is often low. Suggest why.

....

.... **(1 mark)**

Electron shells and orbitals

1 The maximum number of electrons in the first four shells is

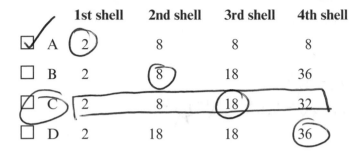

	1st shell	2nd shell	3rd shell	4th shell
☑ A	2	8	8	8
☐ B	2	8	18	36
☐ C	2	8	18	32
☐ D	2	18	18	36

 (1 mark)

2 (a) Draw a sketch of a p-orbital.

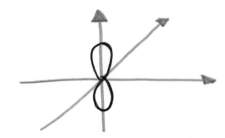

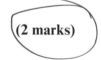 (1 mark)

Guided (b) Why in each p sub-shell is there is a maximum of six electrons?

In the p sub-shell there are ..**3**...... p-orbitals, each of which can contain

...2..electrons...in..atomic..orbital...with..opposite..spins 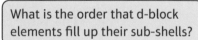(2 marks)

(c) Four sub-shells in the order in which they are filled with electrons are

☐ A   1s 1p 2s 2p

☐ B   2s 2p 3p 3d

☑ C   3s 3p 4s 3d

☐ D   3d 4s 4p 4f

> What is the order that d-block elements fill up their sub-shells?

~~(1 mark)~~

3 This is a diagram of a d-orbital.

Which of the statement(s) are correct?

1   This orbital can contain a maximum of 10 electrons.

2   If the orbital contains more than one electron they must have different spin.

3   Only the second and subsequent energy levels can have a d-orbital.

☑ A   only 1   ☐ B   only 2   ☐ C   only 3   ☐ D   1, 2 and 3

~~(1 mark)~~

Electron configurations – filling the orbitals

1 (a) Give the electron configuration of a

fluorine atom ...$1s^2 2s^2 2p^5$.................................... ①

chlorine atom ...$1s^2 2s^2 2p^6 3s^2 3p^5$.................. ①

bromine atom ...$1s^2 2s^2 2p^6 3s^2 3p^6 4s^2 3d^{10} 4p^6$.... ①   **(3 marks)**

(b) Use your answers to (a) to justify why each of these elements
is placed in Group 7.

> Which sub-shells should you consider?

...All of the outer electrons held in p-sub shell.
...There are 7 electrons in the highest shell...   **(1 mark)**

2 Give the symbol of a particle which has the electronic configuration $1s^2 2s^2 2p^6$ $=10$
and is

(a) neutral .... Ne ....   **(1 mark)**

(b) a cation with a 3+ charge .... Al^{3+} $(13-3=10)$ ....   **(1 mark)**

(c) an anion with a 2− charge .... O^{2-} $(8+2=10)$ ....   **(1 mark)**

3 (a) The electrons-in-boxes electron configuration of an atom is given.

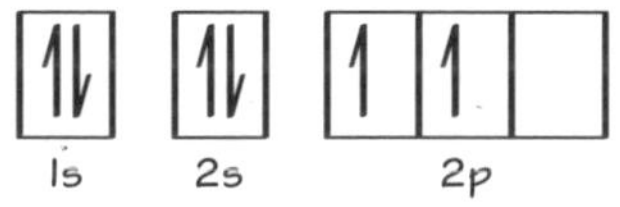

[↑↓] [↑↓] [↑|↑|]
 1s   2s   2p

Which of the statement(s) are correct?

1 This is the electron configuration of a carbon atom.   O

2 This atom has electrons in three shells.   o

3 This is the electron configuration of an N⁻ ion.   o

☐ A only 1

☐ B only 1 and 2

☐ C only 2 and 3

☑ D 1, 2 and 3 $[Ar] 4s^2 3d^{10}$ → $[Ar] 3d^{10}$ **(1 mark)**

Guided (b) Give the electron configuration of

(i) a zinc atom [Ar] ...$1s^2 2s^2 2p^6 3s^2 3p^6 4s^2 3d^{10}$.........   **(1 mark)**

(ii) a Zn^{2+} ion [Ar] ...$1s^2 2s^2 2p^6 3s^2 3p^6 4s^2 3d^8$.........   **(1 mark)**

(iii) a chromium atom, using electrons-in-boxes notation.

[Ar] ...$4s^2 3d$..   **(1 mark)**

$1s^2 2s^2 2p^6 3s^2 3p^6$

Ionic bonding

1   Aluminium sulfide is an ionic compound which exists in several lattice structures.

   (a) Draw, using dots (●) for sulfur electrons and crosses (✖) for aluminium electrons, showing outer shells only,

   (i)   a sulfur ion   2,8,6

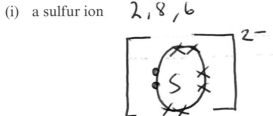

> Remember to show the charge.

(1 mark)

   (ii) a aluminium ion. 2,8,3

(1 mark)

   (b) Deduce the formula of aluminium sulfide.

   .... Al_2S_3 ....   (1 mark)

> **Guided**

   (c) What is meant by a **lattice**?

   A regular arrangement ..of..ions..repeated..many..times..:.........

   ....   (2 marks)

   (d) Aluminium sulfide has a melting point of 1100 °C. Explain why the melting point is high.

   ..Ionic..compound:..Insufficient..energy..at..room..temp..to..overcome

   ..strong..electrostatic..forces..of..attraction..between..opp..charged ions (2 marks)

2   (a) The formula of a compound is X_3Y_2.

   The groups in which elements X and Y are likely to be found are

	X	Y
☐ A	2	7
☐ B	2	5
☐ C	3	6
☐ D	3	4

(1 mark)

   (b) The most likely melting and boiling points of Al_2O_3 are

	$T_m/°C$	$T_{mb}/°C$
☑ A	2070	2980
☐ B	−120	230
☐ C	230	260
☐ D	1200	1080

(1 mark)

Covalent bonds

1 (a) What is a covalent bond?

...Covalent...bond...is...the...strong...electrostatic...force...of...
...attraction...between...a...shared...pair...of...electrons...and...
the nuclei of the bonded atoms.   **(2 marks)**

(b) The diagram below shows a covalent molecule. Oxygen electrons are shown with a dot. Hydrogen and nitrogen electrons are shown by a cross.

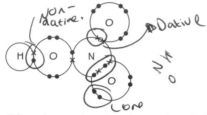

(i) Give the molecular formula of the molecule.

...HNO_3...   **(1 mark)**

(ii) Give the name of the molecule.

...Nitric acid...

> This molecule is often used as a solution.

(1 mark)

(iii) Explain what is meant by a dative covalent bond. On the diagram, circle a dative covalent bond and label it **D** and circle a non-dative covalent bond and label it **N**.

...Dative covalent bond = one atom donates both...
...electrons to be shared...
....   **(3 marks)**

(iv) What is a lone pair? Circle a lone pair on the diagram and label it **L**.

> In which shell are the relevant electrons found?

...Pair of electrons not used in bonding...
....   **(2 marks)**

▷**Guided**▷ 2 Draw a *dot and cross* diagram of a molecule of propene, $CH_3CH=CH_2$, showing outer electrons only, below.

(3 marks)

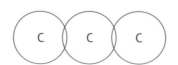

Shapes of molecules

1   The molecule which is planar is

☐ A PH_3

☐ B C_2H_6

☐ C SiH_4

☑ D   B(Cl_3)

> Count the number of electron pairs around P, C, Si and B.

(1 mark)

Maths skills

2 (a) (i) Draw, and state the name of, the shape of a molecule of PF_5.

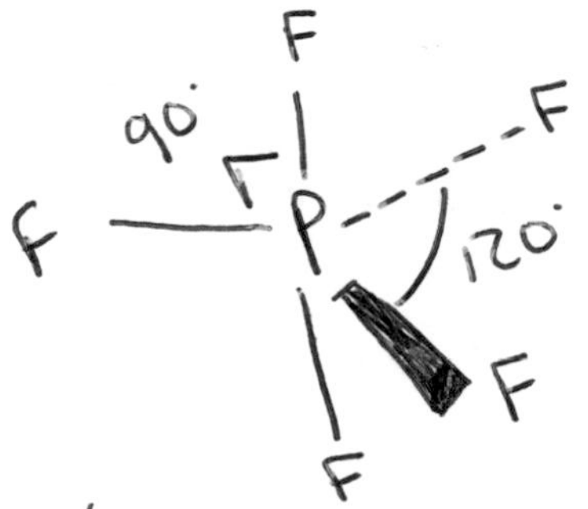

name .... Trigonal bipyramid ....   **(2 marks)**

(ii) Clearly label on your diagram the two different F–P–F bond angles and give the values.   **(2 marks)**

Guided

(iii) Explain why a PF_5 molecule has this shape.

There are .... 5 .... bond pairs around

....P....which..repel..to..be..as..far..apart..as..possible   **(2 marks)**

3 (a) Draw a diagram of a molecule of sulfur trioxide, SO_3, showing all covalent bonds.

> In this molecule, sulfur has 12 electrons in its outer shell.

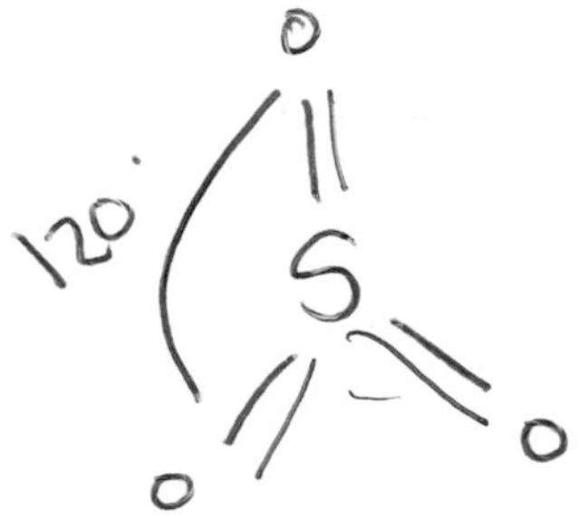

(2 marks)

(b) State the shape and give the value of the O–S–O bond angle in this molecule.

.... Trigonal..planar..and..120°..................   **(2 marks)**

(c) Write an equation for the formation of this compound from sulfur, S_8, and oxygen.

.... S_8 | $12O_2 \longrightarrow 8SO_3$ ....   **(1 mark)**

$S=8$
$O=23$

$S=1$
$O=23$

21

More shapes of molecules and ions

1 The diagrams below show a molecule of methane and a molecule of ammonia.

methane	ammonia
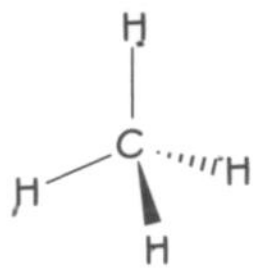	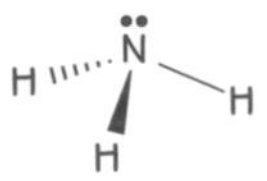

(a) (i) State the shape of the methane molecule.

......Tetrahedral.....

(1 mark)

(ii) Explain why a methane molecule has this shape.

.......- 4 bonding pairs...............................

.....- Lone pairs repel more strongly than..

.....bonding pairs..

> What is repelling to determine the shape of this molecule?

(2 marks)

(iii) Give the value of the H–C–H bond angle.

................109.5°............

(1 mark)

(b) (i) State the shape of the ammonia molecule.

......Pyramidal....

(1 mark)

(ii) State the H–N–H bond angle and explain why this is unlike the H–C–H bond angle in methane.

............107.4...

......- 3 bonding , 1 lone :..........................

.......- lone pair repel more strongly than..........

.....bonded pair :..

(3 marks)

2 The shape of a molecule of H_2S is

☑ A linear

☐ B non-linear

☐ C trigonal planar

☐ D triangular

$H_2 (S_2)$

2 bond
2 lone.

(1 mark)

Electronegativity and bond polarity

1 (a) What is meant by the term **electronegativity**?

Electronegativity is a measure of the attraction of a bonded atom for the pair of electrons in a covalent bond. **(2 marks)**

(b) State and explain the trend in electronegativity going across Period 3.

> Consider the shielding and the nuclear charge.

- Electronegativity increases

....

.... **(3 marks)**

(c) (i) Draw an example of a **polar covalent** bond formed between two atoms in Period 3, showing all partial charges.

(2 marks)

(ii) State and explain whether the molecule NH_3 is polar.

Yes, Polar, dipoles do not cancel out **(3 marks)**

2 Consider a molecule of water.

Which of the statement(s) below are correct?

1 Some ionic compounds dissolve in water because water is a polar molecule.

2 A stream of water from a burette is bent by a charged rod.

3 The polarity of the two OH bonds cancels because they are opposite each other.

☐ A   1, 2 and 3   ☐ C   only 2 and 3

☑ B   only 1 and 2   ☐ D   only 1   **(1 mark)**

> Guided

3 Alcohols consist of a hydrocarbon chain and an O–H group. The solubility of four alcohols is given.

Alcohol	Structure	Solubility/ mol per 100 g water
butan-1-ol	$CH_3CH_2CH_2CH_2OH$	0.11
pentan-1-ol	$CH_3CH_2CH_2CH_2CH_2OH$	0.030
hexan-1-ol	$CH_3CH_2CH_2CH_2CH_2CH_2OH$	0.0058
heptan-1-ol	$CH_3CH_2CH_2CH_2CH_2CH_2CH_2OH$	0.00080

State and explain the trend in solubility by considering bond polarities.

The carbon chain is ......*polar*...... so does not interact with

...*water*... The O–H bond is ...*Polar and causes solubility. Longer*...

...*the carbon chain, less affect of O-H, less soluble alcohol.*... **(3 marks)**

Had a go ☐ Nearly there ☐ Nailed it! ☐

Van der Waals' forces

Guided

1 (a) (i) Explain how interatomic forces arise in a sample of the noble gas neon.

Fluctuations in the atoms' distribution causes a temporary dipole ....

....

.... **(3 marks)**

(ii) The boiling points of the noble gas elements are given. Draw a graph of atomic number of the element against the boiling point on the grid. **(3 marks)**

Element	Boiling point/ K
helium	4
neon	27
argon	84
krypton	122
xenon	167

(iii) State and explain the trend shown in the graph.

....

.... **(3 marks)**

2 In a sample of hydrogen chloride gas, which of the following, per mole, is the strongest?

HCl(g) consists of simple molecules.

☐ A London forces

☐ B Permanent dipole–dipole forces

☑ C Covalent bonds

☐ D Ionic bonds **(1 mark)**

3 Which substance has the highest boiling point?

Do the molecules all have the same surface area? How does this affect the intermolecular forces?

☑ A $CH_3CH_2CH_2CH_2CH_3$ – longest

☐ B $CH_3CH(CH_3)CH_2CH_3$

☐ C $C(CH_3)_4$

☐ D They all have the same boiling point. **(1 mark)**

Hydrogen bonding and the properties of water

1 The boiling point of the hydrides of the first four elements of Groups 4 and 5 are given in the table.

Hydride	Boiling point/ K	Hydride	Boiling point/ K
CH_4	112	NH_3	240
SiH_4	161	PH_3	185
GeH_4	184	AsH_3	218
SnH_4	221	SbH_3	256

(a) Plot on the grid the boiling point of the hydride of each Group 4 element against the period in which the element is found. Join each point to each subsequent point (do **not** draw a best-fit line). Then repeat this on the same axes with the hydrides of the group 5 elements.

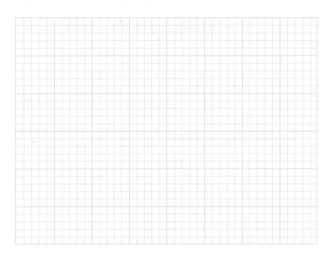

(5 marks)

(b) State and explain the trend seen in the boiling point of the Group 4 hydrides.

....

....

.... **(3 marks)**

(c) Explain why the boiling point of ammonia is relatively high.

....

| What additional bonding is found only in ammonia? |

....

.... **(3 marks)**

25

Properties of simple molecules

1 Some covalent bond enthalpies and melting points are given in the table.

Substance	Bond enthalpy/ kJ mol^{-1}	Melting point/ °C
chlorine	Cl–Cl = +243	−102
diamond	C–C = +347	3527
iodine	I–I = +151	114
water	H–O = +459	100

(a) What is meant by **bond enthalpy**?

....

.... **(2 marks)**

>Guided> (b) Explain why the bond enthalpy of I–I is less than that of Cl–Cl.

The electrostatic attraction between the shared pair and ....

....

.... **(3 marks)**

(c) Explain why the melting point of diamond is very high compared with that of water, even though the C–C bond enthalpy is lower than the H–O bond enthalpy.

....

.... **(2 marks)**

2 Which substance is most likely to have a simple molecular lattice?

		Melting point/ °C	Conductivity of electricity	Reaction with water
☐	A	−7	none	dissolves
☐	B	98	high	reacts
☐	C	2072	none	insoluble
☐	D	3915	high	insoluble

(1 mark)

3 (a) Iodine forms a simple molecular lattice.

Draw a diagram of part of the lattice, showing at least eight iodine atoms.

(2 marks)

(b) When heated iodine has sublimed, the particles present are

☐ A I atoms   ☐ B I_2 molecules   ☐ C I$^-$ ions   ☐ D I$^+$ ions.

(1 mark)

Exam skills 2

1 Boron forms a compound with hydrogen, called borane. It is a colourless gas. Another compound of boron and hydrogen exists, of higher relative molecular mass, called diborane.

(a) Each of these molecules consists of 78.3% boron by mass. Calculate the empirical formula of these molecules.

....

....

.... **(3 marks)**

(b) The relative molecular mass of diborane is 27.6. Give the molecular formula of each gas.

> Use the information at the top of the page.

diborane .... borane .... **(2 marks)**

(c) (i) Draw a *dot and cross* diagram of a molecule of borane.

(2 marks)

(ii) State and explain the shape of a borane molecule, giving the bond angle.

....

....

.... **(3 marks)**

(d) Diborane can be synthesised by reacting boron trifluoride, BF_3, with lithium hydride, LiH. The other product is $LiBF_4$.

(i) Construct the equation for this reaction.

.... **(1 mark)**

(ii) Give the oxidation numbers of lithium and hydrogen in LiH.

Li .... H .... **(2 marks)**

(iii) Explain why lithium hydride has a much higher melting point than diborane.

....

.... **(2 marks)**

The Periodic Table

1 Silicon is an element, with a melting point of 1414°C.

(a) Give the electronic configuration of a silicon atom.

.... **(1 mark)**

(b) In which block of the periodic table is silicon placed?

> Use your answer to part (a) to help.

☐ A d-block

☐ B f-block

☐ C p-block

☐ D s-block **(1 mark)**

(c) In the periodic table, silicon is placed between aluminium and phosphorus.

(i) Explain why these three elements are found in this order.

....

....

.... **(2 marks)**

(ii) Explain why silicon has such a high melting point.

....

> Consider the structure of silicon and the strength of the bonds between silicon atoms.

....

.... **(3 marks)**

⟩Guided⟩ (iii) Explain why the melting point of phosphorus is much lower than that of silicon.

Phosphorus consists of .... molecules ....

....

.... **(2 marks)**

(d) Silicon(IV) oxide reacts with sodium oxide to form sodium silicate, Na_2SiO_3.

> What two types of substance are needed?

Explain why this can be described as a neutralisation reaction.

....

....

.... **(2 marks)**

Ionisation energy

1 (a) Give the definition of **first ionisation energy**.

....

....

.... **(3 marks)**

> The specification lists definitions you should know – this is one of them, so learn it precisely.

(b) Write equations for the reaction where the energy change is

> Apply the definition from (a) and remember to include state symbols.

(i)   the first ionisation of energy of magnesium.

.... **(2 marks)**

(ii) the second ionisation energy of oxygen.

.... **(2 marks)**

2 The first ionisation energy for sodium is $495 \, \text{kJ} \, \text{mol}^{-1}$.

(a) The first ionisation energy for potassium is $420 \, \text{kJ} \, \text{mol}^{-1}$.

Explain why the first ionisation energy for potassium is lower.

....

....

....

.... **(3 marks)**

Maths skills

Guided

(b) Which graph shows the first four ionisation energies of sodium?

☐ A

☐ C

> It gets more difficult to remove an electron because the particle it is being removed from gets more positive.

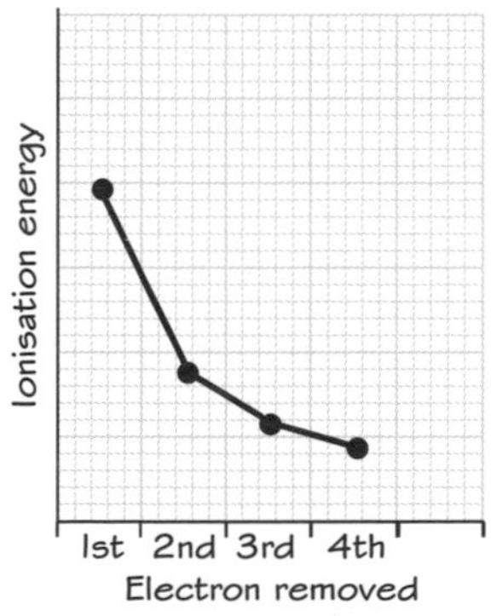

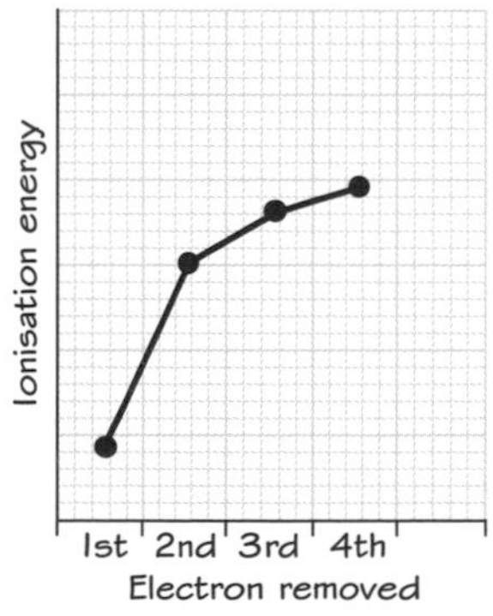

☐ B

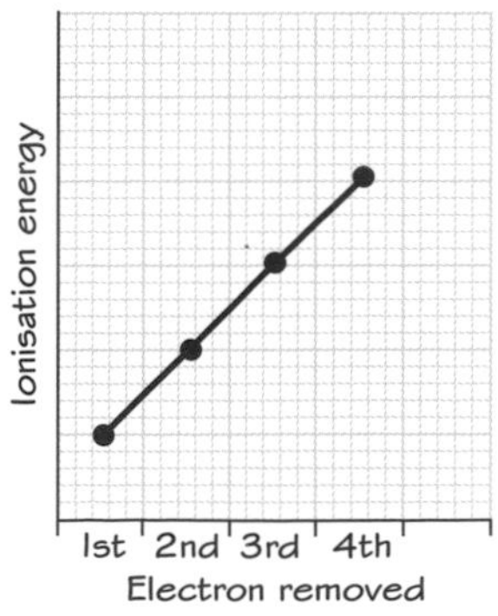

☐ D

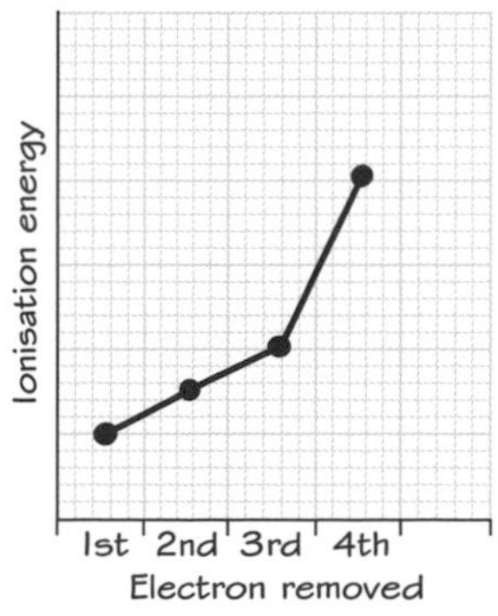

(1 mark)

Ionisation energy across Periods 2 and 3

 Maths skills

1   The graph shows the first ionisation energy of the elements in Period 2.

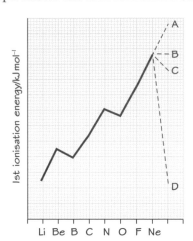

> Think about the group this element is found in, and compare it with the element on the graph in the same group.

(a) Which letter, A, B, C or D, shows the first ionisation energy for the element after Ne?

.... **(1 mark)**

⟩Guided⟩ (b) Explain why the general trend is that the first ionisation energy increases across Period 2.

The first ionisation energy increases across Period 2 because each

subsequent atom has one extra ....

The .... being removed is in the same shell so that

.... from inner electrons is ....

Overall, the .... attraction of the .... to

the nucleus is ...., meaning that ....

energy is required.   **(3 marks)**

(c) (i)   Complete the electronic configuration of an oxygen atom.   **(1 mark)**

⟨↑↓⟩ ⟨↑↓⟩ ⟨ | | ⟩
1s   2s   2p

> What effect does the pairing of two of the 2p electrons have?

(ii) Explain why the first ionisation energy of oxygen is lower than that of nitrogen.

....

....

....

....

....   **(3 marks)**

Structures of the elements

1 Carbon exists in three structures, diamond, graphene and graphite. A diagram of graphene is shown.

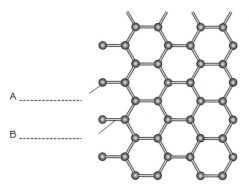

(a) Complete the labels A and B on the diagram.   **(2 marks)**

(b) Draw a diagram to show the structure of diamond.

> Draw sufficient atoms to show the three-dimensional structure of diamond. How many atoms is each carbon atom joined to?

                                                                                        (2 marks)

(c) Another form of carbon is buckminsterfullerene, which consists of C_{60} molecules.

What is the structure of the three forms of carbon in (a), (b) and (c)?

	Graphene	Diamond	Buckminsterfullerene
☐ A	giant covalent	giant covalent	giant covalent
☐ B	simple molecular	giant covalent	simple molecular
☐ C	giant covalent	giant covalent	simple molecular
☐ D	simple molecular	simple molecular	simple molecular

                                                                                        (1 mark)

2 Argon, chlorine, fluorine, neon, nitrogen, oxygen, phosphorus and sulfur are all elements in Periods 2 and 3 with low boiling points.

(a) Put the elements phosphorus, sulfur, chlorine and argon in order of increasing boiling point.

....   **(1 mark)**

Guided

(b) Explain why all of these elements have relatively low boiling points.

These elements all exist as separate atoms or small ....

....

....   **(3 marks)**

31

Properties of the elements

Guided 1 Join with a line each of the Period 3 elements with its description at
room temperature.

sodium   •   • a yellow solid

magnesium   •   • element with the highest boiling point

aluminium   •   • structure consists of separate atoms

silicon   •   • a silver solid that reacts vigorously with water

phosphorus   •   • a green gas

sulfur   •   • element with the highest electrical conductivity

chlorine   •   • a silver solid that reacts slowly with water

argon   •   • structure consists of molecules with four atoms

(8 marks)

2 Explain why aluminium is an excellent conductor of electricity,
silicon conducts electricity slightly and sulfur is a non-conductor
of electricity.

> What is required in the structure of an element for electricity to be conducted? How many of these particles do the elements have?

....

....

....

....

....

....

.... **(4 marks)**

3 Magnesium reacts with water to form magnesium hydroxide and a gas.

(a) Write the equation for this reaction

.... **(2 marks)**

(b) In an experiment the pH of the solution formed when some magnesium
was reacted with water was 10, but when repeated with sodium the pH
of the solution formed was 13.

> Which ion causes an alkaline pH?

Explain these observations.

....

....

.... **(3 marks)**

Group 2 elements

1 Calcium reacts with water.

   (a) Write the equation for this reaction.

   .... **(2 marks)**

   (b) Describe what observations would be made when a small piece of calcium is added to a beaker of water.

> **Practical skills**   Think about the state of each of the products that you have included in the equation.

   ....

   .... **(2 marks)**

   (c) The product of the reaction described in (b) is filtered and the filtrate divided into two portions.

      (i) Describe and explain the observations that would be made if a few drops of universal indicator are added to a portion.

      ....

      .... **(2 marks)**

Guided

      (ii) A student blows through a straw into a portion to test for one of their exhaled gases.

         The solution formed in (b) is known as ....

         Exhaled breath contains the gas ....

         which gives .... **(2 marks)**

2 Magnesium reacts with hydrochloric acid.

   (a) Write the equation for this reaction.

   .... **(2 marks)**

   (b) This is a redox reaction.

      (i) What is a **redox** reaction?

      ....

      .... **(1 mark)**

      (ii) Explain in terms of electrons how magnesium changes, and describe this change in terms of redox.

> Have the magnesium atoms gained or lost electrons?

      ....

      .... **(2 marks)**

Group 2 compounds and their uses

1   Group 2 oxides react with water.

(a)   In an experiment, a piece of calcium oxide is added to a test-tube of water and some universal indicator is added.

> **Practical skills** You should be able to describe reactions that are mentioned in the specification.

Explain the colour change that would be seen.

....

.... **(2 marks)**

(b)   The reaction of strontium oxide with water is exothermic.

(i)   Suggest what observation would suggest that this reaction is exothermic.

> Which observation depends on the reaction mixture getting hot?

....

.... **(2 marks)**

> **Guided**

(ii)   Write the equation for the reaction of strontium oxide with water.

SrO + ... **(2 marks)**

(c)   When water is added to some magnesium oxide there is no observable change.

The pH of a mixture of a small amount of magnesium oxide with a test tube of water is most likely to be

> The reaction is slight ('no observable change') but which ions are formed?

☐   A   5

☐   B   7

☐   C   9

☐   D   13 **(1 mark)**

> **Practical skills**

2   Tablets containing calcium carbonate are used to treat acid indigestion. These tablets also contain inert material.

You are given:
- a bottle containing twenty tablets, mass 500 mg
- dilute hydrochloric acid, 0.500 mol dm^{-3}
- usual laboratory apparatus
- indicators

> **Practical skills** Give the essential practical details, including the apparatus, the measurements and an outline of the results analysis.

Describe how you would carry out an experiment to determinate the percentage by mass of calcium carbonate in the tablets.

....

....

....

.... **(4 marks)**

The halogens and their uses

1 (a) Complete the table below giving some information about the halogens.

	M_r	Colour	Boiling point/ °C
fluorine		pale yellow	−188
chlorine	71	green	−34
bromine	160		59
iodine	254	black	184

(2 marks)

(b) State and explain the trend in the boiling points in the halogens.

....

....

.... **(3 marks)**

2 Chlorine reacts with sodium hydroxide solution forming sodium chloride, sodium chlorate(I), NaClO, and water.

(a) Write the equation for this reaction.

.... **(2 marks)**

(b) Give the oxidation number of **chlorine** in each of the species in the equation:

Cl_2 ....

NaCl ....

NaClO .... **(3 marks)**

⟩Guided⟩ (c) Using your answers in (b), state and explain the **type** of reaction that occurs when sodium hydroxide reacts with chlorine.

Chlorine as an element has an oxidation number ....

which .... to .... in NaCl and ....

.... to .... in NaClO. A reaction where one substance

is simultaneously .... and ....is called a

.... reaction. **(2 marks)**

(d) Chlorine and its compounds have many uses.

Which answer is an **incorrect** statement about chlorine and drinking water?

☐ A   Chlorine reacts with organic matter in the water to form compounds which may be carcinogenic.

☐ B   Chlorine is safe to use in drinking water because it is not toxic to humans.

☐ C   Chlorine kills bacteria in the drinking water.

☐ D   Chlorine reacts when added to the drinking water to form chloric(I) acid. **(1 mark)**

Reactivity of the halogens

1 (a) Give the electronic configuration of

    (i) a fluorine atom .... **(1 mark)**

    (ii) a fluoride ion .... **(1 mark)**

(b) Iron reacts with fluorine to form iron(III) fluoride.

    (i) Write the equation for this reaction.

    .... **(2 marks)**

    (ii) In this reaction, iron is

    ☐ A displaced

    ☐ B neutralised

    ☐ C oxidised

    ☐ D reduced. **(1 mark)**

    (iii) Chlorine, bromine and iodine also react with iron.

    Describe and explain the pattern in reactivity that you would expect to be demonstrated in these reactions. You are **not** expected to give any observations.

> No experimental details are expected here, only a statement of whether the halogens get more or less reactive (with iron) and the reason why (in terms of the halogen atoms).

    ....

    ....

    ....

    .... **(3 marks)**

⟩Guided⟩ 2 In an experiment to test the reactivity of halogens, halogen solutions are added to solutions of sodium halides. The results are:

	Cl^- (aq)	Br^- (aq)	I^- (aq)
Cl_2 (aq)		orange solution	brown solution
Br_2 (aq)	orange solution		brown solution
I_2 (aq)	brown solution	brown solution	

> **Practical skills** In three of these, the colour is due to a halogen product. In the other three, it is the colour of the original, unreacted halogen.

Analyse these results, explaining them using the order of reactivity of the halogens.

The brown solution when chlorine and bromine are added to iodide ion

shows the formation of ....

....

.... **(3 marks)**

Tests for ions

1 The test for carbon dioxide is to bubble the gas through limewater.

(a) Write the equation for this reaction.

.... **(2 marks)**

(b) (i) Give the observation that you would make.

.... **(1 mark)**

(ii) Give the name of the substance that gives rise to the observation in (i).

.... **(1 mark)**

> The name is required – you must not give the formula. Remember, though, that you have used the formula in (a).

2 You are given some pure, white crystals containing two ions.

(a) Sodium hydroxide solution is added to a sample of the crystals and the mixture warmed. A gas is evolved which turned damp red litmus paper blue.

The gas evolved is

☐ A   ammonia

☐ B   carbon dioxide

☐ C   sulfur dioxide

☐ D   ammonium.   **(1 mark)**

> **Guided**

(b) (i) Describe how you would show that the crystals contain sulfate ions.

The steps in the method are:

1. Add the sample to a test tube and ....

2. To the solution, add .... then ....

3. A .... forms, indicating sulfate ions.   **(3 marks)**

(ii) Write the ionic equation for this test.

.... **(2 marks)**

> Show only the ions that lead to the formation of the precipitate.

(c) Using the information in (a) and (b), give the formula of the crystals.

.... **(1 mark)**

> In contrast to 1(b)(ii), here you must give the formula, not the name.

3 Some silver halides dissolve in dilute and/or concentrated ammonia solutions.

Which of the following statements is correct?

☐ A   Silver chloride is insoluble in dilute ammonia.

☐ B   Silver bromide is soluble in dilute ammonia but insoluble in concentrated ammonia.

☐ C   Silver bromide is soluble in dilute ammonia and concentrated ammonia.

☐ D   Silver iodide is insoluble in ammonia solution.   **(1 mark)**

Exam skills 3

1 Bromine is the only non-metal element that is liquid at room temperature and pressure.

(a) Using the trends in Group 7, suggest the colour and state of fluorine and of astatine at room temperature and pressure.

....

.... **(3 marks)**

(b) (i) Bromine is extracted from natural bromine-rich deposits. It was the first element to be extracted from seawater, but this is now only economically viable at the Dead Sea.

> 'Suggest' means that you have not been taught this directly, but use the information given and/or your other knowledge.

Suggest why the extraction of bromine from seawater is only viable in one location.

....

.... **(1 mark)**

(ii) A solution of a pure bromide salt is tested to show that bromide ions are present. Describe this test, including the use of ammonia.

....

....

.... **(3 marks)**

(iii) Explain why the test above would **not** be reliable when carried out on seawater.

....

.... **(2 marks)**

(c) (i) Write an equation for the reaction whose enthalpy change is the first ionisation energy of bromine.

.... **(2 marks)**

(ii) Explain why the first ionisation energy of krypton is higher than that of bromine.

....

.... **(2 marks)**

(d) Bromine has an oxidation number of +3 in

☐ A KBrO

☐ B NaBr

☐ C $KBrO_3$

☐ D BrF_3. **(1 mark)**

Enthalpy profile diagrams

1   The diagram below shows an enthalpy profile diagram.

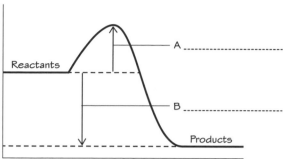

  (a)   Complete the diagram with the labels A and B.   **(2 marks)**

Guided   (b)   Explain how you know that this diagram is for an exothermic reaction.

The enthalpy of the reactants is ....

....

.... **(1 mark)**

  (c)   Draw a labelled enthalpy profile diagram for an endothermic reaction.

            Adapt the diagram given above.

(2 marks)

2 (a)   What is the meaning of the term **activation energy**?

....

.... **(2 marks)**

  (b)   One reason that some reactions happen slowly is that:

     ☐   A   only a small proportion of reactant particles have the activation energy

     ☐   B   the average energy of the reactant particles is less than the activation energy

     ☐   C   reactions can only happen if heat is released

     ☐   D   the activation energy is higher than the enthalpy change.   **(1 mark)**

Enthalpy change of reaction

1 The apparatus that can be used to measure the molar enthalpy change of solution of some crystals is shown.

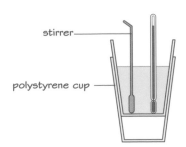

stirrer

polystyrene cup

The method a student used was

• Measure 50 cm³ distilled water with a measuring cylinder.

• Transfer the water to the polystyrene cup, and take the temperature.

• Weigh the crystals, then add to the water and stir until completely dissolved.

• Take the final temperature.

(a) What is meant by the **molar enthalpy change of solution**?

....

.... **(2 marks)**

(b) Explain why a polystyrene cup is used rather than a glass beaker.

.... **(1 mark)**

 Practical skills You are trying to measure the temperature change accurately. There is more than one improvement possible.

(c) Suggest and explain one improvement that could be made to the apparatus.

....

.... **(2 marks)**

(d) Suggest and explain one improvement that could be made to the procedure, if the apparatus remains the same.

....

.... **(2 marks)**

(e) For measurements at standard conditions, the pressure must be

☐ A   100 Pa

☐ B   1000 Pa

☐ C   100 000 Pa

☐ D   101 325 Pa (1 atmosphere). **(1 mark)**

Calculating enthalpy change

 Maths skills

1 An experiment was carried out to determine the enthalpy change of the reaction between magnesium and copper sulfate.

$$Mg(s) + CuSO_4(aq) \rightarrow MgSO_4(aq) + Cu(s)$$

The data for the experiment is

mass of magnesium/ g	2.45
volume of copper sulfate solution/ cm^3	50.0
concentration of copper sulfate solution/ mol dm^{-3}	2.00
initial temperature/ °C	18.4
final temperature/ °C	64.2

> **Maths skills** The significant figures in the data tell you how many significant figures you should use in your final answer.

(a) Calculate the amount of heat evolved in kJ. ($c = 4.18 \, \text{J K}^{-1} \text{g}^{-1}$)

> The mass used in the equation $q = mc\Delta T$ is the mass of the copper sulfate solution **only**.

....

....

....

.... $q = $ .... kJ   **(3 marks)**

(b) (i) Calculate the moles of magnesium.

.... moles = ... mol   **(1 mark)**

(ii) Calculate the moles of copper sulfate.

.... moles = ... mol   **(1 mark)**

(c) Calculate the enthalpy change of reaction in kJ mol^{-1}.

> Remember to give the correct sign. Is this an exothermic or endothermic reaction?

....

$\Delta_r H = $ .... kJ mol^{-1} **(2 marks)**

> **Guided**

(d) The data book value for this reaction is more exothermic.

Which of the reason(s) below could explain this?

1   Heat was lost to the environment

2   The copper sulfate solution was more dilute than stated

~~3~~   An excess of magnesium was added

> This makes no difference as all of the copper sulfate has reacted.

☐ A   1, 2 and 3

☐ B   only 1 and 2

☐ C   only 2 and 3

☐ D   only 1   **(1 mark)**

Enthalpy change of neutralisation

1 A student carries out an experiment to determine the enthalpy change of neutralisation.

$50.0\,cm^3$ of $2.00\,mol\,dm^{-3}$ NaOH is placed in a container.

$70.0\,cm^3$ of hydrochloric acid is added in $5.00\,cm^3$ portions.

After each addition, the temperature is taken.

Results

Vol acid / cm^3	Temp. / °C	Vol acid / cm^3	Temp. / °C	Vol acid / cm^3	Temp. / °C
0.0	23.5	25.0	29.9	50.0	36.4
5.0	24.8	30.0	31.3	55.0	35.2
10.0	26.2	35.0	32.6	60.0	34.0
15.0	27.4	40.0	34.0	65.0	32.8
20.0	28.7	45.0	35.2	70.0	31.6

(a) Draw a graph of the results of the experiment. **(4 marks)**

Guided

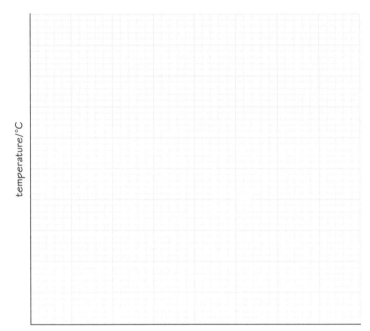

temperature/°C

volume of acid added/cm3

> Choose your scale so that at least half of the graph paper is used, with an even scale on each axis. Draw **two** best-fit lines on this graph – one as temperature rises and one as temperature falls.

(b) Use your graph to find the temperature rise in the experiment. Show your working.

> Where do the best-fit lines cross?

....

.... temperature rise = ...°C **(2 marks)**

(c) If the volume of acid required to neutralise the sodium hydroxide is $V\,cm^3$, and your answer to (b) is $T°C$, the heat evolved, q, in J, is

☐ A $120 \times 4.18 \times T$

☐ B $50 \times 4.18 \times T$

☐ C $(50 + V) \times 4.18 \times T$

☐ D $(120/1000) \times 4.18 \times T$ **(1 mark)**

Hess' law

1 Iron is produced by the reduction of iron(III) oxide, Fe_2O_3, with carbon monoxide.

(a) Write the equation for this reaction.

.... **(1 mark)**

Maths skills

(b) A Hess' law cycle including this reaction and three others is shown.

$$3Fe_2O_3 + 9CO \xrightarrow{3\Delta_r H^\ominus} 6Fe + 9CO_2$$

$$\downarrow \qquad\qquad \uparrow$$

$$2Fe_3O_4 + CO_2 + 8CO \implies 6FeO + 3CO_2 + 6CO$$

Use the cycle and the standard enthalpy changes of reaction below to calculate the standard enthalpy change of the reaction in (a), shown by $\Delta_r H^\ominus$.

> Match each arrow on the diagram with the equations given and fill in the data on the diagram.
> - One is reversed (so reverse the sign of $\Delta H^\ominus$).
> - Multiply the data as appropriate.
> - Then calculate the answer by adding the three values.

$3Fe_2O_3(s) + CO(g) \rightarrow 2Fe_3O_4(s) + CO_2(g)$ $\Delta H^\ominus = -48.3\,kJ\,mol^{-1}$

$Fe(s) + CO_2(g) \rightarrow FeO(s) + CO(g)$ $\Delta H^\ominus = +10.9\,kJ\,mol^{-1}$

$Fe_3O_4(s) + CO(g) \rightarrow 3FeO(s) + CO_2(g)$ $\Delta H^\ominus = +21.8\,kJ\,mol^{-1}$

....

....

.....................................$\Delta_r H^\ominus =$ ....$kJ\,mol^{-1}$ **(3 marks)**

Guided

(c) Give one condition for the Hess' law cycle to be valid.

The temperature of each step included in the cycle ....

.... **(1 mark)**

(d) The enthalpy change for the reaction in (a) cannot be measured directly because

☐   A   it is impossible to get pure carbon monoxide gas

☐   B   the reaction can only be carried out using a catalyst

☐   C   it is impossible to get a high enough temperature for the reaction to occur

☐   D   iron(III) oxide reacts with carbon monoxide in two different reactions.   **(1 mark)**

Enthalpy change of formation

1 Ammonium nitrate, NH_4NO_3, is an ionic solid.

> You must include state symbols.

Write an equation for which the enthalpy change is
the enthalpy of formation of ammonium nitrate.

.... **(3 marks)**

2 Consider the equation below, for which the enthalpy change is **X**.

$2H_2(g) + O_2(g) \rightarrow 2H_2O(g)$

The standard enthalpy of formation of water

☐ A   is **X**

☐ B   is 2**X**

☐ C   is $\dfrac{\mathbf{X}}{2}$

☐ D   can be calculated using **X** and one more piece of data.   **(1 mark)**

Maths skills

3 Glucose, $C_6H_{12}O_6$, can be fully combusted in oxygen.

(a) Write the equation for the complete combustion of one mole of glucose,
including state symbols, with all four substances in standard states at 298 K.

.... **(2 marks)**

Guided

(b) Calculate $\Delta_r H^\ominus$ for the reaction given the following enthalpy of formation data.

	$\Delta_f H^\ominus/\text{kJ mol}^{-1}$
$C_6H_{12}O_6(s)$	−1275
$CO_2(g)$	−394
$H_2O(l)$	−286

> No data is given for oxygen, because it is an element so $\Delta_f H^\ominus = 0$.

$\Delta_r H^\ominus = \sum \Delta_f H^\ominus \text{ (products)} - \sum \Delta_f H^\ominus \text{ (reactants)}$ ....

....

....

....$\Delta_r H^\ominus =$ ....kJ mol^{-1} **(3 marks)**

(c) Sucrose, $C_{12}H_{22}O_{11}$, can be dehydrated using concentrated sulfuric acid.

$C_{12}H_{22}O_{11} \rightarrow 12\,C + 11\,H_2O$

Calculate $\Delta_r H^\ominus$ for the reaction given that the enthalpy of formation of
sucrose is −2223 kJ mol^{-1}, and the data in (ii).

....

....

....

....$\Delta_r H^\ominus =$ ....kJ mol^{-1} **(3 marks)**

Enthalpy change of combustion

1 The standard enthalpy change of combustion of ethane is the enthalpy change at 298 K for the reaction

☐ A $C_2H_6(g) + \frac{7}{2}O_2(g) \rightarrow 2CO_2(g) + 3H_2O(g)$

☐ B $C_2H_6(g) + \frac{7}{2}O_2(g) \rightarrow 2CO_2(g) + 3H_2O(l)$

☐ C $2C_2H_6(g) + 7O_2(g) \rightarrow 4CO_2(g) + 6H_2O(g)$

☐ D $2C_2H_6(g) + 7O_2(g) \rightarrow 4CO_2(g) + 6H_2O(l)$ **(1 mark)**

Practical skills

2 (a) The liquid alcohol, propan-1-ol, burns readily in air.

> The energy released on combustion is used to heat up water.

Describe in outline a practical method to find the enthalpy of combustion of propan-1-ol.

Describe the apparatus and measurements. Do not give any details about calculations.

....

....

....   **(4 marks)**

Guided

(b) (i) In such an experiment as the one described in (a), the mass of propan-1-ol burnt was 0.900 g. This heated 100 g water by 54.4 °C.

Calculate the enthalpy of combustion of propan-1-ol. ($c = 4.18\,J\,K^{-1}\,g^{-1}$)

> **Maths skills** The $q = mc\Delta T$ equation gives an answer in J. Convert to kJ by dividing by 1000.

$q = mc\Delta T =$ ....

....

....

moles of propan-1-ol = ...

....$\Delta_c H^{\ominus} =$ ....$kJ\,mol^{-1}$ **(3 marks)**

(ii) The data book value for the standard enthalpy of combustion of propan-1-ol is $-2021\,kJ\,mol^{-1}$.

Calculate the % error in the result from (b) (i).

.... error = ...%   **(1 mark)**

Maths skills

3 Calculate the standard enthalpy change of reaction for $6C(s) + 7H_2(g) \rightarrow C_6H_{14}(l)$ given the standard enthalpies of combustion, in $kJ\,mol^{-1}$.

C(s)	−394	H₂(g)	−286	C₆H₁₄(l)	−4163

....

....

....$\Delta_r H^{\ominus} =$ .... $kJ\,mol^{-1}$ **(3 marks)**

Bond enthalpies

1 Some bond enthalpies are given in the table.

	Bond	Bond enthalpy / kJ mol^{-1}
1	H–H	432
2	H–Br	364
3	Cl–Cl	243
4	Br–Br	193
5	I–I	149

	Bond	Average bond enthalpy / kJ mol^{-1}
6	C–C	347
7	C=C	614
8	C–H	413
9	C–Br	276

(a) Write an equation for the reaction for which the enthalpy change is 432 kJ mol^{-1}.

.... **(1 mark)**

(b) Explain why values 1–5 are bond enthalpies, but values 6–9 are **average** bond enthalpies.

....

.... **(2 marks)**

(c) The reason for the trend in values 3–5 is that going down Group 7

☐ A   the relative formula mass of the molecule increases

☐ B   the van der Waals' forces between the molecules become stronger

☐ C   the bond length increases

☐ D   the elements get less reactive. **(1 mark)**

Maths skills

Guided

(d) Use the data to calculate the enthalpy change of reaction for

$CH_3CH=CH_2(g) + Br_2(l) \rightarrow CH_3CHBrCH_2Br(g)$

Bond enthalpies in reactants = ...

....

Bond enthalpies in products = ...

....$\Delta_r H^\ominus$ = ...kJ mol^{-1} **(3 marks)**

Maths skills

(e) Use the data to calculate the enthalpy change for

$C(g) + 4Br(g) \rightarrow CBr_4(g)$

....$\Delta_r H^\ominus$ = ...kJ mol^{-1} **(1 mark)**

Collision theory

 1 Zinc reacts with dilute hydrochloric acid. An experiment is carried out to find the effect of altering the concentration of the acid on the rate of reaction, using the apparatus shown.

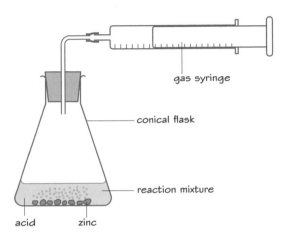

gas syringe

conical flask

reaction mixture

acid   zinc

Acid is measured into the flask. A weighed amount of zinc in lumps is added to the acid and the stopper inserted. The volume of gas in the syringe is measured at regular intervals until the reaction is over.

(a) Which gas is evolved in the reaction?

.... **(1 mark)**

(b) How would the experimenter know that the reaction was over?

.... **(1 mark)**

(c) If a scientist is provided with dilute hydrochloric acid of concentration $1.0\,mol\,dm^{-3}$, $50\,cm^3$ of dilute hydrochloric acid of concentration $0.25\,mol\,dm^{-3}$ can be made by mixing:

> **Practical skills** How many times more dilute do you want the acid to be?

☐ A $40\,cm^3$ dilute hydrochloric acid and $10\,cm^3$ distilled water

☐ B $0.25\,dm^3$ dilute hydrochloric acid and $1\,dm^3$ distilled water

☐ C $12.5\,cm^3$ dilute hydrochloric acid and $37.5\,cm^3$ distilled water

☐ D $10\,cm^3$ dilute hydrochloric acid and $40\,cm^3$ distilled water.   **(1 mark)**

(d) Explain, in terms of particles, why the reaction is faster if the acid is more concentrated.

....

....

(2 marks)

Guided (e) By considering the method and apparatus, state why the amount of gas collected will be lower than that evolved in the reaction, and suggest a method of avoiding this error.

The reaction begins when ....

....

....   **(3 marks)**

Measuring reaction rates

Maths skills

1 In a reaction the concentration of a reactant, A, is measured at regular intervals.
The data is given below.

Time/min	[A]/mol dm^{-3}	Time/min	[A]/mol dm^{-3}	Time/min	[A]/mol dm^{-3}
0	0.500	4	0.305	8	0.219
1	0.431	5	0.278	9	0.205
2	0.379	6	0.255	10	0.192
3	0.338	7	0.236	11	0.181

(a) Plot a graph of the data on the grid below. **(4 marks)**

> **Maths skills** Do not forget to label your axes and give units. Draw a best-fit curve.

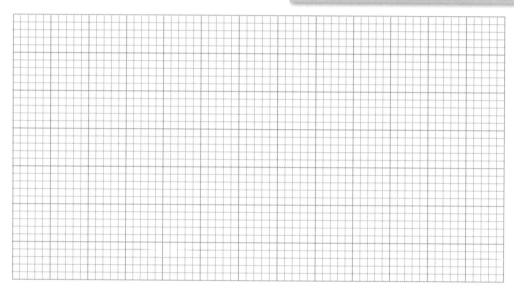

Guided (b) Use the graph to calculate the **initial** rate of reaction.

Initial rate = gradient = $\dfrac{\Delta y}{\Delta x}$

....

.... initial rate = ... mol dm^{-3} min^{-1} **(3 marks)**

Practical skills (c) Explain whether the reaction had finished at 11 minutes.

....

.... **(2 marks)**

(d) The concentration of reactants A and B and
product C in a reaction are plotted.

The most likely equation for the reaction is

☐ A   A + B → C

☐ B   2A + B → C

☐ C   A + 2B → C

☐ D   A + B → 2C

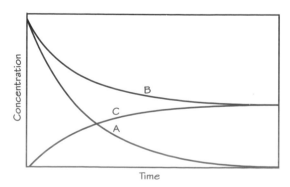

(1 mark)

The Boltzmann distribution

1 The Boltzmann distribution shows the distribution of kinetic energy amongst particles in a gas.

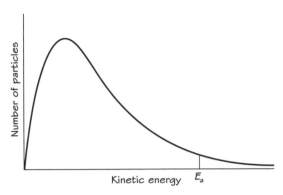

(a) Use the diagram to state why only a small proportion of collisions are successful.

....

.... **(1 mark)**

(b) Draw on the graph a second line showing the distribution of energy for the same amount of gas but at a lower temperature.   **(2 marks)**

(c) Use the diagram to explain the effect on the rate of reaction of reducing the temperature.

What proportion of the particles have energy $\geqslant E_a$?

....

....

....

.... **(3 marks)**

⟩**Guided**⟩ (d) Which of the statement(s) about the Boltzmann distribution is correct?

When the curve is drawn for the same sample of gas at a higher temperature,

1   the most likely kinetic energy increases

2   the number of particles with the most likely kinetic energy increases

3̶   the activation energy increases.

The activation energy is the same whatever the conditions.

☐ A   1, 2 and 3

☐ B   only 1 and 2

☐ C   only 2 and 3

☐ D   only 1

Catalysts

1 Hydrogen peroxide decomposes.

$2H_2O_2(aq) \rightarrow 2H_2O(l) + O_2(g)$

This reaction can be catalysed by some drops of solution containing iodide ions.

(a) An experiment was carried out testing whether a solution of sodium iodide acted as a catalyst for the decomposition. It was found that sodium iodide solution **did** act as a catalyst.

Design an experiment to show that the iodide ions, and **not** the sodium ions, acted as the catalyst.

1. A measured volume of hydrogen peroxide solution was placed in a flask with a gas syringe connected and left for a period of time.

2. The experiment was repeated, but some drops of ....

....

3. The experiment was repeated again, but some drops of ....

4. Only the reaction with .... added was faster

showing that ....

.... **(4 marks)**

(b) The catalysed reaction occurs in two steps.

The first step is $H_2O_2 + I^- \rightarrow H_2O + IO^-$

The equation for the second step is

> The two steps must combine to make the overall equation, and the catalyst must be regenerated in step 2.

☐ A $H_2O + IO^- \rightarrow 2H_2O + O_2$

☐ B $H_2O_2 + I^- \rightarrow H_2O + O_2$

☐ C $H_2O_2 + IO^- \rightarrow H_2O + O_2 + I^-$

☐ D $H_2O_2 + I^- \rightarrow H_2O + O_2 + I_2$ **(1 mark)**

(c) Catalysts can be homogeneous or heterogeneous.

Explain the difference between these two types of catalyst, and state which type of catalyst the iodide ions are in the decomposition.

....

.... **(2 marks)**

Dynamic equilibrium

1 When ethanol is mixed with ethanoic acid in the presence of a concentrated sulfuric acid catalyst and left, a dynamic equilibrium is reached.

$$C_2H_5OH + CH_3COOH \rightleftharpoons CH_3COOC_2H_5 + H_2O$$

(a) Explain two features of a **dynamic equilibrium**.

....

.... **(2 marks)**

(b) The concentration of ethanoic acid is monitored over time.

The results are shown on the graph.

> **Practical skills**   In (iii) outline how to take samples, stop the reaction in the sample, and measure the concentration of acid in the sample.

(i) Add to the graph a line showing the concentration of $CH_3COOC_2H_5$ over time. **(2 marks)**

(ii) Mark on the graph the time when dynamic equilibrium has been reached. **(1 mark)**

(iii) Outline a way that you could measure the concentration of the ethanoic acid in the laboratory.

....

....

.... **(4 marks)**

(c) Which of the statement(s) about the equilibrium is correct?

   1   At equilibrium the rates of the forward and the backward reaction are equal.

   2   At equilibrium the concentrations of all the substances are equal.

   3   At equilibrium the amounts of product are higher if a catalyst is added.

☐   A   1, 2 and 3

☐   B   only 1 and 2

☐   C   only 2 and 3

☐   D   only 1   **(1 mark)**

Guided   (d) Suggest a safety precaution when the reaction is being set up.

The .... is corrosive, so ....

.... **(1 mark)**

Le Chatelier's principle

Guided

1 State Le Chatelier's principle.

Le Chatelier's principle states that when any change is made to the

....

position of the equilibrium moves in the direction that ....

.... **(2 marks)**

2 Consider the equilibrium $N_2O_4(g) \rightleftharpoons 2NO_2(g)$. $\Delta H = 57\,kJ\,mol^{-1}$.

State and explain the effect on the position of equilibrium of

(a) increasing the temperature.

....

.... **(2 marks)**

(b) increasing the pressure.

....

.... **(2 marks)**

3 Consider the equilibrium $Fe^{3+}(aq) + SCN^-(aq) \rightleftharpoons FeSCN^{2+}(aq)$.

The mixture on the left is pale yellow, the $FeSCN^{2+}$ ion is deep red.

(a) When the equilibrium mixture is cooled, the colour becomes deeper red.

Deduce and explain whether the reaction is exothermic or endothermic.

> In which direction has the position of equilibrium moved? This is the direction which minimises the original cooling.

....

....

.... **(3 marks)**

(b) AgSCN is an insoluble compound. Use Le Chatelier's principle to explain the effect on the position of equilibrium of adding some $AgNO_3(aq)$.

> A precipitate will remove ions from the equilibrium.

....

.... **(2 marks)**

4 In $CO(g) + NO_2(g) \rightleftharpoons CO_2(g) + NO(g)$, where the forward reaction is exothermic, the yield of NO can be increased by

☐ A   reducing the concentration of carbon monoxide

☐ B   reducing the temperature

☐ C   reducing the pressure

☐ D   adding a suitable catalyst. **(1 mark)**

The equilibrium constant

1 For the equilibrium $2SO_2(g) + O_2(g) \rightleftharpoons 2SO_3(g)$, the forward reaction is exothermic.

(a) Write the expression for the equilibrium constant, K_c.

.... **(1 mark)**

(b) How can the value of K_c be increased?

> The question is not asking about yield, but about K_c value.

☐   A   Increasing the concentration of sulfur dioxide

☐   B   Reducing the temperature

☐   C   Increasing the pressure

☐   D   Adding a suitable catalyst   **(1 mark)**

(c) In industry this reaction is carried out at a temperature of around 450 °C.

Explain why this temperature is chosen, and not a higher or lower temperature.

....

....

....

....

.... **(4 marks)**

Maths skills

Guided

(d) In a simulation, under certain conditions the equilibrium mixture was found to have 0.5 mol SO_2, 0.25 mol O_2 and 1.5 mol SO_3 in a vessel of volume 100 cm³.

Use your expression from (a) to find K_c and give the units.

The concentrations of each gas in mol dm⁻³ are ....

Substituting into the K_c expression, K_c = ...

Substituting mol dm⁻³ into K_c to give units ....

.... **(3 marks)**

2 In the reaction $N_2 + 3H_2 \rightleftharpoons 2NH_3$, $K_c = 0.105 \, dm^6 mol^{-2}$ under certain conditions.

This value of K_c indicates that

☐   A   the concentration of NH_3 is 0.105 mol dm⁻³

☐   B   the concentration of NH_3 is $\sqrt{0.105}$ mol dm⁻³

☐   C   the equilibrium position lies to the left hand side

☐   D   10.5% of the mixture is ammonia.   **(1 mark)**

Exam skills 4

Maths skills

1 NOCl decomposes to form nitrogen monoxide and chlorine. $K_c = 1.6 \times 10^{-5}\,\text{mol}\,\text{dm}^{-3}$.

$2NOCl(g) \rightleftharpoons 2NO(g) + Cl_2(g)$

(a) (i) Calculate the enthalpy change of reaction given the following enthalpy of formation data.

	$\Delta_f H^\ominus$ / kJ mol^{-1}
NOCl	51.2
NO	90.3

....

....$\Delta_r H^\ominus =$ ....kJ mol^{-1} **(3 marks)**

(ii) Why is the enthalpy of formation of $Cl_2(g)$ zero?

....   **(1 mark)**

(b) Calculate the enthalpy change of reaction given the following bond enthalpies. Assume that the bonding is O=N–Cl in NOCl and N=O in NO.

	Bond enthalpy/ kJ mol^{-1}
N=O	481
N–Cl	159
Cl–Cl	243

....

....

....$\Delta_r H^\ominus =$ ....kJ mol^{-1} **(3 marks)**

(c) Complete the table with 'increased', 'decreased' or 'unchanged' to give the effect of the change given on the **rate of attainment of equilibrium** and on the **yield of chlorine.**   **(3 marks)**

Change	Effect on rate of attainment of equilibrium	Effect on yield of chlorine
increase of temperature		
increase of pressure		
addition of a catalyst		

(d) The value of K_c for $2NO(g) + Cl_2(g) \rightleftharpoons 2NOCl(g)$ is

☐ A $1.6 \times 10^{-5}\,\text{mol}^{-1}\,\text{dm}^3$

☐ B $1.6 \times 10^{-5}\,\text{mol}\,\text{dm}^{-3}$

☐ C $62\,500\,\text{mol}^{-1}\,\text{dm}^3$

☐ D $62\,500\,\text{mol}\,\text{dm}^{-3}$

> The reaction has been reversed, so the values of the concentrations in K_c will now be inverted.

(1 mark)

Key terms in organic chemistry

1 When chlorine gas is exposed to ultraviolet light, the following reaction occurs.

$Cl_2 \rightarrow 2Cl^{\cdot}$

(a) In this reaction

1   homolytic fission has occurred

2   heterolytic fission has occurred

3   chlorine atoms have formed

Which of the statement(s) are correct?

☐ A   only 1 and 3

☐ B   only 2 and 3

☐ C   only 1

☐ D   only 2   **(1 mark)**

(b) Chlorine radicals are formed. What is a radical?

....   **(1 mark)**

2 Two molecules are shown.

(a) Explain whether **each** molecule is

   (i) alicyclic

   ....

   ....   **(2 marks)**

   (ii) saturated

   ....

   ....   **(2 marks)**

(b) Give the name of molecule **A**.

....   **(1 mark)**

Naming hydrocarbons

1 (a) Give the name of the molecules whose structure is shown.

> How many carbons in the longest chain? What is joined to the middle carbon?

Structure   Name   Structure   Name

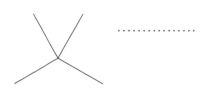

....

....

(4 marks)

(b) When the alcohol $CH_3CH_2CH(OH)CH_3$ is dehydrated, a mixture of but-1-ene and but-2-ene is formed.

   (i)   Write the equation for the formation of one of these molecules.

   .... **(2 marks)**

   (ii) But-1-ene and but-2-ene can be described as

      1   isomers

      2   saturated

      3   alkenes

Which of the statement(s) are correct?

☐ A   only 1 and 2

☐ B   only 1 and 3

☐ C   only 2 and 3

☐ D   1, 2 and 3   **(1 mark)**

   (iii) Draw the structure and name a saturated molecule with the same molecular formula as but-1-ene.

> The molecule has a ring.

(2 marks)

56

Naming compounds with functional groups

1 Consider the molecule shown.

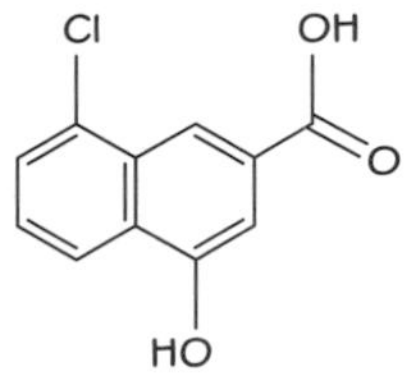

   (a) Identify all of the functional groups in this molecule. Do not include the rings.

   ....

   .... **(3 marks)**

   (b) State the molecular formula of the molecule.

   .... **(1 mark)**

2 A diagram of a molecule is given.

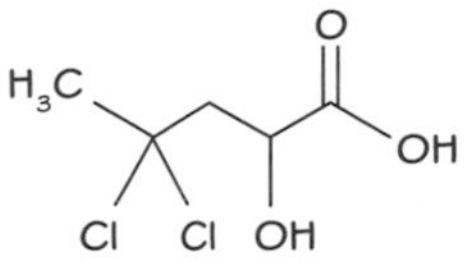

   The name of this molecule is

   ☐ A 4,4-dichloro-2-ol-pentanoic acid

   ☐ B 2,2-dichloro-4-hydroxypentanoic acid

   ☐ C 4,4-dichloro-1,2-dihydroxypentanal

   ☐ D 4,4-dichloro-2-hydroxypentanoic acid. **(1 mark)**

3 Draw the structure of

   (a) 1,1,2-trichloroethane

   (1 mark)

   (b) four molecules with the formula C_3H_7OBr and name each one.

   ....

   .... **(8 marks)**

Different types of formulae

1   Formulae of five molecules are given below.

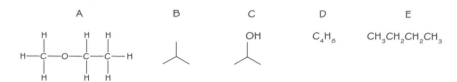

List from the molecules, **A**, **B**, **C**, **D** and **E**

(a) A structural formula ....   **(1 mark)**

(b) A displayed formula ....   **(1 mark)**

(c) A skeletal formula ....   **(1 mark)**

(d) All sets of isomers ....   **(2 marks)**

(e) All hydrocarbons ....   **(1 mark)**

(f) All saturated molecules ....   **(1 mark)**

(g) Give the general formula of the series including
     molecule **B** ....   **(1 mark)**

(h) Draw and name the isomers of molecule **D**. Do not
     include both the *E*- and *Z*-isomers of the same structure.

> There are six marks
> here so you are likely
> to need three isomers.

(6 marks)

2   A molecule containing carbon, hydrogen and oxygen contains 40.0% carbon and
    53.3% oxygen by mass.

(a) (i) Calculate the empirical formula of the molecule

         The percentage of hydrogen = 100 – ....

         ....

         ....   **(3 marks)**

    (ii) The relative molecular mass of the molecule is 60. Write down the molecular formula.

         ....   **(1 mark)**

    (iii) Draw and name one possible structure.

                                              ....   **(2 marks)**

Structural isomers

Guided   **1** (a) Give the definition of structural isomers.

Molecules with the same .... but different ....

.... **(2 marks)**

(b) (i) Draw the displayed formula of propan-1-ol.

(1 mark)

(ii) Draw, and name, a structural isomer of propan-1-ol.

.... **(8 marks)**

2 A saturated molecule containing seven carbons is shown.

(a) Give the molecular formula of this molecule.

.... **(1 mark)**

(b) The IUPAC name of this molecule is

☐ A   2-chloro-cycloheptane

☐ B   1-chloromethylcyclohexane

☐ C   chloromethylbenzene

☐ D   1-chloro-1-methylcyclohexane. **(1 mark)**

(c) Draw the skeletal formula of an isomer of the molecule shown that does not
contain a ring. **(1 mark)**

What type of molecule would have the
correct number of carbons and hydrogens?

Had a go ☐ Nearly there ☐ Nailed it! ☐

Properties and reactivity of alkanes

1 Butane and methylpropane are alkanes. The boiling point of butane is −1 °C, and the boiling point of methylpropane is −11.7 °C.

(a) **(i)** Draw the displayed formula of methylpropane.

(1 mark)

(ii) An example of a molecule that does **not** have a formula matching the general formula for alkanes is

1 methylpropane

2 benzene

3 cyclobutane

☐ A only 1

☐ B only 2

☐ C only 3

☐ D only 2 and 3 **(1 mark)**

Guided

(iii) How can a molecule be identified as an alkane from its structural formula?

The molecule contains only ....

.... **(2 marks)**

(iv) Explain the difference in boiling point between butane and methylpropane.

Consider the intermolecular forces.

....

....

.... **(2 marks)**

(b) Alkanes are generally not very reactive because

☐ A the bonds in alkanes are non-polar

☐ B C–H bonds are weak

☐ C alkanes are inflammable

☐ D carbon and hydrogen atoms in alkanes have full outer shells. **(1 mark)**

Reactions of alkanes

1 Propane reacts with chlorine.

(a) Write the overall equation showing the formation of 1-chloropropane from propane and chlorine.

.... **(2 marks)**

(b) (i) The mechanism for this reaction involves initiation, propagation and termination.

Write the mechanism for the formation of 1-chloropropane showing these three stages.

(4 marks)

(ii) State the name of this mechanism.

.... **(2 marks)**

(c) Ultraviolet light is needed for this reaction to occur. Explain what the light does.

....

.... **(2 marks)**

(d) In the reaction between propane and chlorine, which of the following substances can be formed?

1   1,2-dichloropropane

2   chloromethane

3   hexane

☐ A   only 1 and 2

☐ B   only 1 and 3

☐ C   only 2 and 3

☐ D   1, 2 and 3

Consider the mechanism you drew in (b) (i) and the possible termination steps.

(1 mark)

Bonding in alkenes

Guided   1   The diagram below shows the atoms and some of the covalent bonds in a molecule of ethene.

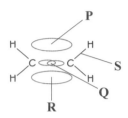

(a) **(i)**   Explain the difference between a σ bond and a π bond.

In a σ bond, the shared pair of ....

....

.... **(3 marks)**

(ii)   In the diagram above, letter(s) indicating a σ bond(s) are

☐ A   P and R

☐ B   Q

☐ C   S

☐ D   Q and S.   **(1 mark)**

(iii)   In the diagram above, letter(s) indicating a π bond are

☐ A   P and R

☐ B   Q

☐ C   S

☐ D   Q and S.   **(1 mark)**

(b)   Butene, an unsaturated molecule, is an alkene.

(i)   What is meant by **unsaturated**?

.... **(1 mark)**

(ii)   How, from its structure, can a molecule be identified as an alkene?

.... **(1 mark)**

(iii)   Draw the displayed formula of a molecule of but-1-ene.

(2 marks)

Stereoisomerism in alkenes

1 (a) What are stereoisomers?

....

.... **(2 marks)**

(b) Consider the molecules below.

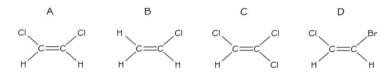

    A   B   C   D

(i) State which of the molecules **A**, **B**, **C** and **D** are able to exist as stereoisomers, and explain why they can exist as stereoisomers.

> There are two marks for the explanation, so make two separate points.

Molecules that exist as stereoisomers ....

Explanation ....

.... **(3 marks)**

(ii) Give the name of molecule **A** using

the *E–Z* naming system. ....

the *cis–trans* naming system .... **(3 marks)**

2 The molecule with the formula C_5H_{10} can exist as isomers.

Draw the displayed formula of

(a) two molecules with molecular formula C_5H_{10} that are *E–Z* isomers

(2 marks)

(b) a straight chain molecule with molecular formula C_5H_{10} that does not have an *E–Z* isomer

(1 mark)

⟩Guided⟩ (c) a molecule with molecular formula C_5H_{10} that is **not** an alkene.

(1 mark)

Addition reactions of alkenes

1 Ethene can be converted to ethanol.

(a) Write the balanced equation for the reaction.

.... **(2 marks)**

(b) Give **two** conditions for this reaction.

.... **(2 marks)**

>Guided> (c) Explain why this reaction is described as an **addition** reaction.

In the addition reaction, a .... is broken

.... **(2 marks)**

2 A partial mechanism is shown for the reaction of ethene with bromine.

(a) Complete the mechanism by adding the missing dipole,
the missing curly arrow and the missing charge on the carbocation. **(3 marks)**

(b) The mechanism of this reaction is

☐ **A** electrophilic substitution

☐ **B** electrophilic addition

☐ **C** electrophilic fission

☐ **D** nucleophilic bromination. **(1 mark)**

(c) Some bromine water is shaken with some cyclohexane and with some
cyclohexene. State what would be observed in each case.

....

....

.... **(2 marks)**

Formation and disposal of polymers

1 The diagram below was taken from a student's notebook, showing the polymer poly(styrene). The hexagonal symbol has the formula C_6H_5.

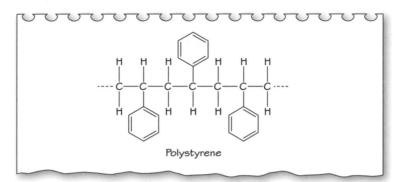

Polystyrene

(a) (i) Draw the structure of the monomer that would produce poly(styrene).

(1 mark)

(ii) Describe how monomers form a polymer molecule.

....

.... **(2 marks)**

(b) The empirical formula of poly(styrene) is

☐ A $(CH_2–CHC_6H_5)$

☐ B $(CH_2–CHC_6H_5)_n$

☐ C $(CH_2=CHC_6H_5)$

☐ D CH **(1 mark)**

2 The polymer PVC is formed from the monomer with the structure shown.

Cl

(a) Give the IUPAC name of PVC.

.... **(1 mark)**

(b) Describe some benefits and drawbacks of the disposal of PVC by combustion.

....

....

.... **(4 marks)**

Had a go ☐ Nearly there ☐ Nailed it! ☐

Exam skills 5

1 (a) Draw the displayed formula of a molecule of 1,2-dibromoethane

(1 mark)

(b) A high yield of this compound could be formed by reacting

☐ A ethane and bromine solution

☐ B ethene and bromine

☐ C 1,2-dichloroethane and bromine

☐ D ethane and excess bromine with UV light. **(1 mark)**

(c) Draw and name the displayed formula of a structural isomer of 1,2-dibromoethane.

(2 marks)

2 A molecule, **P**, of formula C_4H_8 reacts with steam and phosphoric acid to form molecule **Q**. Molecule **P** exhibits geometric isomerism.

(a) To which homologous series does **P** belong?

.... **(1 mark)**

(b) To which homologous series does **Q** belong?

.... **(1 mark)**

(c) Draw the displayed formula of one of the geometric isomers of a molecule of **P**.

(1 mark)

(d) Draw the **skeletal** formula of an isomer of **P** that does **not** react with bromine solution.

(1 mark)

(e) Draw the displayed formula of a molecule of an isomer of **P** that reacts with bromine solution but does **not** exhibit geometric isomerism.

(1 mark)

The properties of alcohols

Guided

1 The table gives the boiling points of some alkanes and some alcohols.

Molecule	Boiling point/ °C
methane	−164
ethane	−89
propane	−42
methanol	65
ethanol	79
propan-1-ol	97
propan-2-ol	82

(a) State and explain the trend in the boiling point of the alkanes.

From methane to propane, the boiling points ....

This is because molecules have more ....

.... **(2 marks)**

(b) Compared to alkanes of a similar relative molecular mass, alcohols have a higher

    1   volatility

    2   boiling point

    3   solubility in water

> The polarity of the O–H bond causes the difference in properties.

Which of the statement(s) are correct?

☐   A   only 1

☐   B   only 1 and 2

☐   C   only 2 and 3

☐   D   1, 2 and 3   **(1 mark)**

(c) Draw the displayed formula of a molecule of propan-1-ol and a molecule of propan-2-ol, and suggest why propan-1-ol has a higher boiling point.

....

.... **(4 marks)**

(d) From the alcohols in the table, give the names of **all** of the molecules, stating 'none' if there are none, that are

    (i)   primary alcohols ....

    (ii)   secondary alcohols ....

    (iii) tertiary alcohols ....   **(3 marks)**

Combustion and oxidation of alcohols

Practical skills

1 A student has a sample of three alcohols, **P**, **Q** and **R**, each with the formula $C_5H_{11}OH$. Each alcohol has some potassium dichromate(VI) and dilute sulfuric acid added, and the mixtures are warmed.

(a) Alcohol **Q** shows no colour change.

(i)   What can be deduced about the structure of alcohol **Q**?

....   **(1 mark)**

(ii) Give the structural formula of alcohol **Q**.

....   1 mark)

(b) Alcohols **P** and **R** show a colour change.

(i)   What colour change is seen?

....   **(2 marks)**

(ii) What can be deduced about the structures of alcohols **P** and **R**?

....   **(1 mark)**

(iii) What is the function of the mixture of potassium dichromate(VI) and dilute sulfuric acid?

....   **(1 mark)**

Practical skills

2 The apparatus shown can be used to change a mixture of an alcohol, potassium dichromate(VI) and dilute sulfuric acid into an aldehyde.

What would happen to the aldehyde if heated under reflux? Explain why this apparatus is used to form an aldehyde, rather than heating the mixture under reflux.

....

....

....   **(2 marks)**

More reactions of alcohols

Guided

1 Pentan-2-ol can be converted to an alkene.

(a) What type of reaction is this?

☐ A   substitution

☒ B   addition

> A double bond is in the reactant for addition.

☐ C   oxidation

☐ D   elimination.   **(1 mark)**

(b) Give the catalyst and condition for this reaction.

.... **(2 marks)**

(c) Give the structures and names of two alkene products that are structural isomers.

> You do not need to consider stereoisomers.

name 1 ....   name 2 ....   **(2 marks)**

2 An alcohol, **X**, is converted to form the haloalkane shown below.

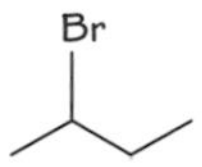

(a) Give the skeletal formula of **X.**

                                                                                (1 mark)

(b) What type of reaction is this?

☐ A   substitution

☐ B   addition

☐ C   oxidation

☐ D   elimination.   **(1 mark)**

(c) Give the reagents needed for the conversion.

.... **(2 marks)**

Nucleophilic substitution reactions of haloalkanes

Guided   1 Propene can be converted to propane-1,2-diol in two steps.

$$\underset{\text{step 1}}{} \qquad \underset{\text{step 2}}{}$$

$$CH_2{=}CH{-}CH_3 \quad \rightarrow \quad CH_2BrCHBrCH_3 \quad \rightarrow \quad \text{propane-1,2-diol}$$

(a) Give the name of the product of step 1.

1,2- .... **(1 mark)**

(b) Draw the displayed formula of propane-1,2-diol.

(1 mark)

(c) The types of reaction in each step are

	step 1	**step 2**
☐ A	addition	elimination
☐ B	addition	substitution
☐ C	substitution	addition
☐ D	substitution	elimination

(1 mark)

(d) Give the reagent required for each step.

step 1 .... **(1 mark)**

step 2 .... **(1 mark)**

2 Samples of 1-chlorobutane, 1-bromobutane and 1-iodobutane are mixed with a mixture of silver nitrate solution and ethanol. Precipitates form in different amounts of time: the first to form is in the 1-iodobutane reaction and the last in the 1-chlorobutane reaction.

(a) Name and give the colour of the precipitate that forms in the 1-iodobutane reaction.

.... **(2 marks)**

(b) Suggest why ethanol is added to the mixture.

.... | Ethanol is **not** a reactant, but a solvent. |

(1 mark)

(c) Explain the order of formation of the precipitates.

.... **(1 mark)**

Preparing a liquid haloalkane

1 The apparatus shown can be used to prepare a haloalkane from an alcohol, a sodium halide and sulfuric acid.

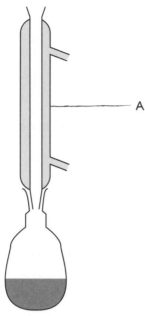

A

(a) Explain why an electrical heater would be used, rather than a Bunsen burner.

.... **(1 mark)**

(b) Name apparatus **A** and explain its purpose.

....

.... **(2 marks)**

(c) State whether the water flows in to the upper or lower opening of apparatus **A**, and explain why.

....

.... **(2 marks)**

(d) The reaction mixture is purified using the steps below.

| Transfer reaction mixture to separating funnel and add washing agent | Shake funnel gently, releasing pressure as necessary | Remove unwanted layer. Transfer organic layer to flask, add drying agent | Filter off drying agent, then distil off the product |

Guided

(i) Explain why in this reaction a suitable liquid for washing is sodium hydrogencarbonate solution.

This is an alkali ....

.... **(1 mark)**

(ii) Give the name of a suitable drying agent.

....

.... **(1 mark)**

Haloalkanes in the environment

1 The skeletal formula of a CFC is shown.

$$\underset{Cl}{\overset{Cl}{\Big\backslash}}\underset{Cl}{\overset{}{\diagup}}F$$

(a) Give the name of the molecule.

.... **(1 mark)**

(b) Give the molecular formula of the molecule.

.... **(1 mark)**

> Remember that each carbon atom forms four bonds.

(c) If the molecule is exposed to UV radiation, free radicals can form.

> Guided

(i) What is a **free radical**?

An atom or group of atoms with .... **(1 mark)**

(ii) Give the **formula** of the most likely free radical that would form from this molecule.

.... **(1 mark)**

(iii) The reason that hydrogen free radicals do not form is that

☐ A the C–H bond is non-polar

☐ B the C–H bond is too strong

☐ C the C–H bond does not vibrate when exposed to radiation

☐ D the C–H bond uses hydrogen's only electron. **(1 mark)**

(d) (i) Describe, using equations, how this CFC molecule catalyses the breakdown of ozone.

....

....

....

....

.... **(4 marks)**

(ii) Explain why this ozone breakdown is harmful.

....

....

....

.... **(2 marks)**

Organic synthesis

1 Butan-2-ol could be formed from butane in two steps.

(a) In the first step, butane is reacted with chlorine.

   (i) What reaction condition is necessary for this step?

   ....   **(1 mark)**

   (ii) Explain why this reaction can be classified as free radical substitution.

   ....

   ....   **(2 marks)**

> Identify the free radical and for what it substitutes.

   (iii) Give the structure of all possible isomers of formula C_4H_9Cl that could be formed in this reaction. Do not include both forms of an optical isomer in your answer.

                                                                                    (2 marks)

   (iv) Write the balanced equation for the formation of one of the isomers that you have given in (b) (iii).

   ....   **(2 marks)**

(b) In the second step C_4H_9Cl is used to form butan-2-ol. Give the IUPAC name of the isomer that can form butan-2-ol.

   ....   **(1 mark)**

(c) Butan-2-ol could be formed in one step by reacting but-1-ene with a reagent in the presence of a catalyst.

   (i) Name the reagent that reacts with the but-1-ene.

   ....   **(1 mark)**

   (ii) The catalyst used is

   ☐ A   phosphoric acid

   ☐ B   sodium hydroxide

   ☐ C   potassium dichromate(VI)

   ☐ D   nickel.   **(1 mark)**

> Guided

   (iii) In this reaction, butan-1-ol and butan-2-ol are formed. Explain which of these will be the major product.

   Markownikoff's rule predicts that .... is the major

   product. This forms via a .... carbocation ....   **(2 marks)**

Exam skills 6

Practical skills

1 A molecule **Z**, $C_6H_{12}Br_2$, can be prepared from bromine and a straight chain alkene.

(a) Describe the colour change that you would see when the reaction occurs.

.... **(1 mark)**

(b) The reaction produces a mixture containing **Z**, unreacted liquid alkene and some aqueous substances.

   (i) How should the organic and non-organic substances be separated?

   ☐ A filtration

   ☐ B distillation

   ☐ C using a Buchner funnel

   ☐ D using a separating funnel. **(1 mark)**

   (ii) When most of the aqueous substances have been removed, a drying agent is added to absorb the remaining water. Give the:

   name of a suitable drying agent .... **(1 mark)**

   observation you would make when the mixture is shaken with a drying agent

   .... **(1 mark)**

   method of removing drying agent .... **(1 mark)**

   (iii) When the organic mixture is dry, how can **Z** be separated from other organic substances?

   .... **(1 mark)**

(c) (i) Draw the **three** possible position isomers of **Z**.

(3 marks)

   (ii) The alkene from which **Z** is formed does **not** exhibit geometric isomers. Name the alkene and explain why it does not exhibit geometric isomerism.

   ....

   .... **(2 marks)**

Infrared spectroscopy

Use the data to help you answer the questions.

Bond	Location	Wavenumber / cm⁻¹
C–C	Alkanes, alkyl chains	750–1100
C–O	Alcohols, esters, carboxylic acids	1000–1300
C=O	Aldehydes, ketones, carboxylic acids, esters	1630–1820
C–H	Alkyl groups, alkenes, arenes	2850–3100
O–H	Carboxylic acids	2500–3300 (broad)
O–H	Alcohols, phenols	3200–3600

1 (a) When molecules are exposed to infrared radiation, the change that gives rise
       to an infrared spectrum is caused by

   ☐ A   the molecules gaining kinetic energy and moving faster

   ☐ B   electrons gaining energy and moving to higher energy levels

   ☐ C   bonds in the molecules gaining energy and vibrating more

   ☐ D   the molecules releasing heat energy.   **(1 mark)**

   (b) The IR spectrum below is of a compound whose molecules contain two
        carbon atoms and one oxygen atom.

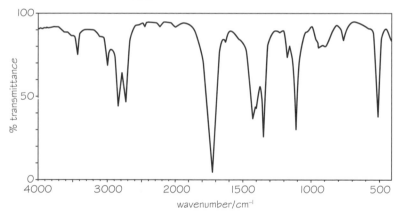

   (i)   Give the structures of two molecules that contain two carbon atoms and
         one oxygen atom.

                                                                        (2 marks)

Guided

   (ii) Use the spectrum to identify the compound, giving your reasons.

         There is a peak at .... cm⁻¹ which shows ....

         but no broad peak at ....cm⁻¹ which shows ....

         so the substance is ....   **(3 marks)**

Uses of infrared spectroscopy

Use the data on the previous page to help answer the questions.

1 (a) Which of the following is **not** a greenhouse gas?

☐ A Carbon dioxide   ☐ C Water vapour

☐ B Nitrogen   ☐ D Methane   **(1 mark)**

Guided   (b) Explain how increased levels of greenhouse gases lead to global warming.

When exposed to IR radiation, bonds in the molecules ....

....

.... **(3 marks)**

(c) In an experiment some ethanol has been oxidised under different conditions, and the products purified. The IR spectra of the two products are given.

(i) What is the oxidising agent used?

.... **(1 mark)**

(ii) Name the two possible oxidation products of the oxidation of ethanol, and identify which product is formed in **A** and which in **B**, with reasons.

> One product is formed by partial oxidation.

....

....

....

....

.... **(4 marks)**

Mass spectrometry

1 The mass spectrum of a molecule, containing carbon, hydrogen and oxygen only, is shown.

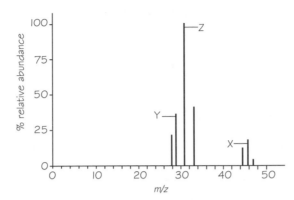

(a) (i) On the *x*-axis, labelled *m*/*z*, what is represented by *m* and *z*?

m .... **(1 mark)**

z .... **(1 mark)**

(ii) What is the name of the species that produces the peak labelled X?

.... **(1 mark)**

(b) (i) What is the molecular mass of the molecule?

.... **(1 mark)**

(ii) Explain the origin of the very small peak that is just to the right of the peak labelled X.

....

.... **(2 marks)**

(c) (i) Suggest the formula of a species, containing **no** oxygen, leading to the peak labelled Y.

.... **(1 mark)**

(ii) Suggest the formula of a species, containing carbon, hydrogen and oxygen, leading to the peak labelled Z.

.... **(1 mark)**

(iii) Use your answers to (b)(i), (c)(i) and (c)(ii) to suggest an identity for the molecule, by name or formula.

.... **(1 mark)**

Concentration–time graphs (zero order reactants)

1 Iodine reacts with propanone under acidic conditions.

$$CH_3COCH_3 + I_2 \mapsto CH_3COCH_2I + HI$$

An experiment was carried out to find the order with respect to iodine.

The method was:

1. $25.0\,cm^3$ propanone and $25.0\,cm^3$ $1.00\,mol\,dm^{-3}$ dilute sulfuric acid are mixed in a flask

2. Add $50.0\,cm^3$ $0.02\,mol\,dm^{-3}$ iodine solution, start timing, and swirl

3. Every 5 minutes, take a $10.0\,cm^3$ sample of the reaction mixture

4. Add sample to $10.0\,cm^3$ $0.500\,cm^3$ sodium hydrogencarbonate solution

5. Titrate the sample, to find the iodine concentration, using $0.0100\,mol\,dm^{-3}$ sodium thiosulfate solution

(a) Why is sulfuric acid added in step 1?

.... **(1 mark)**

> The acid does not appear in the equation.

(b) What should be used to measure the volume of iodine solution in step 2?

> You need to add an accurate volume quickly.

.... **(1 mark)**

 Guided   (c) Explain why the sample is added to sodium hydrogencarbonate solution in step 4.

The sodium hydrogencarbonate reacts with ....

This is called ....

.... **(2 marks)**

(d) The reaction is zero order with respect to iodine.

The graph of the concentration of iodine against time is

☐ A
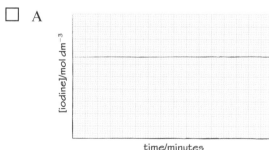

☐ B
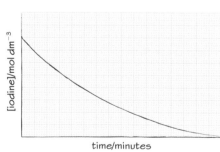

☐ C
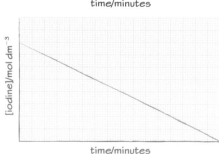

☐ D
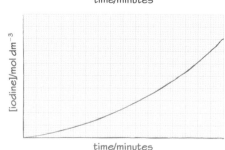

(1 mark)

Concentration–time graphs (first order reactants)

Maths skills

1 In a decomposition reaction, the concentration of the reactant, **X**, is measured over time.

Time/ s	[X]/ mol dm^{-3}
0	0.100
100	0.069
200	0.048
300	0.034
400	0.025
500	0.018

(a) Plot the data on the graph and draw a line of best fit.

(4 marks)

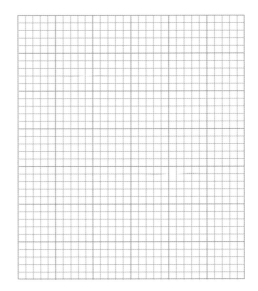

> Draw a line from exactly half of original concentration (= 0.050) to line, and then down to the time axis.

(b) Calculate the half-life starting from

   (i)   time = 0 s .... half-life = ...s   **(1 mark)**

   (ii) time = 200 s .... half-life = ...s   **(1 mark)**

(c) Use your results from (b) to explain the order with respect to **X**.

> How do the half-lives (approximately) compare?

....

....   **(2 marks)**

2 In a decomposition reaction, the original composition of the reactant, **Y**, is 1.00 mol dm^{-3}. Which of the following statements is correct?

   ☐ A   If the reaction is zero-order, [**Y**] is constant.

   ☐ B   If the reaction is first-order, [**Y**] = 0 after two half-lives.

   ☐ C   If the reaction is first-order, [**Y**] = 0.125 mol dm^{-3} after three half-lives.

   ☐ D   If the reaction is first-order and [**Y**] = 0.25 mol dm^{-3} after 4 minutes, $t\frac{1}{2}$ = 1 minute.   **(1 mark)**

Rate equation and rate constant

1. For the reaction $2NO + Cl_2 \rightarrow 2NOCl$, the rate equation is

   rate $= k[NO]^2[Cl_2]$

   (a) State the order with respect to NO, Cl_2 and the overall order.

   NO: .... Cl_2: .... overall: ....   **(2 marks)**

   (b) When $[NO] = 0.13\,mol\,dm^{-3}$ and $[Cl_2] = 0.20\,mol\,dm^{-3}$, the initial rate of reaction is $0.010\,mol\,dm^{-3}\,s^{-1}$.

      (i) Calculate k and give the units.

      > **Maths skills** The units given for the data at the start of part (b) will help you.

      $k = $ ....   **(3 marks)**

      (ii) Use your value of k to find the initial rate when $[NO] = [Cl_2] = 2.00\,mol\,dm^{-3}$.

      rate = ...

      ....   **(2 marks)**

   > Guided

   (c) Explain why the **initial** rate is given in the data for (b).

   At the start the concentrations of the reactants ....,

   but as the reaction proceeds ....

   ....   **(2 marks)**

2. The rate equation for the one-step reaction of **A** with **B**, in the presence of an acid catalyst H^+, to form product **C** is rate $= k[A]^2[B][H^+]$

   Which of the equation(s) below are consistent with this information?

   1. $2A + B \rightarrow C$
   2. $A + 2B \rightarrow C$
   3. $2A + B + H^+ \rightarrow C$

   ☐ A   1, 2 and 3

   ☐ B   only 1

   ☐ C   only 1 and 2

   ☐ D   only 1 and 3

      (1 mark)

Finding the order

1 A student carries out an experiment to determine the orders of reactants.

The equation is A + 2B → 2C, and an acid catalyst, H^+, is present.

(a) Each of the reactants is varied in turn, and the results are shown.

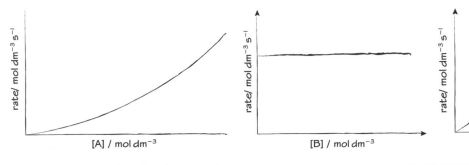

The rate equation for the reaction is

☐ A   rate = $k[A][B]^2$

☐ B   rate = $k[A][B]^2[H^+]$

☐ C   rate = $k[A][H^+]$

☐ D   rate = $k[A]^2[H^+]$

> Look at the shape of the graph for each reactant – is it a horizontal line, a linear graph or a quadratic graph?

(1 mark)

(b) When the experiment is carried out to find the change in rate when the concentration of H^+ catalyst is varied, A and B are used in large excess.

> If **A** and **B** were not in large excess, what would happen to the rate as they get used up?

Explain why a **large excess** is used.

....

....

.... **(3 marks)**

2 The following experimental data were found for the reaction involving **A**, **B** and **C**.

Experiment	[A]/ mol dm^{-3}	[B]/ mol dm^{-3}	[C]/ mol dm^{-3}	Initial rate/ mol dm^{-3} s^{-1}
1	1.25×10^{-3}	1.25×10^{-3}	1.25×10^{-3}	0.0087
2	2.50×10^{-3}	1.25×10^{-3}	1.25×10^{-3}	0.0174
3	1.25×10^{-3}	2.50×10^{-3}	1.25×10^{-3}	0.0348
4	1.25×10^{-3}	3.02×10^{-3}	3.75×10^{-3}	0.457

Determine the order with respect to **A**, **B** and **C**, and the overall order.

....

....

A .... order, **B** .... order, **C** .... order,

overall order = ... **(4 marks)**

Had a go ☐ Nearly there ☐ Nailed it! ☐

The rate-determining step

1 Haloalkanes react with hydroxide ions to form alcohols.

The mechanism for 2-bromo-2-methylpropane's reaction is

Step 1: $(CH_3)_3CBr \rightleftharpoons [(CH_3)_3C]^+ + Br^-$

Step 2: $[(CH_3)_3C]^+ + OH^- \rightleftharpoons (CH_3)_3COH$

(a) Write the complete, overall equation for the reaction of 2-bromo-2-methylpropane with sodium hydroxide.

.... **(1 mark)**

(b) The rate-determining step is step 1.

(i) Write the rate equation for the reaction.

.... **(1 mark)**

(ii) Suggest why step 1, rather than step 2, is the rate-determining step.

> Consider the bonds that have to be broken for each step to occur, and the charges on the reacting particles.

....

.... **(2 marks)**

(c) In the reaction between bromoethane and hydroxide ions, the rate equation is

rate = $k[CH_3CH_2Br][OH^-]$

(i) What can be deduced about the mechanism from this rate equation?

....

.... **(2 marks)**

(ii) Write a mechanism consistent with your answer to part (i).

(3 marks)

(iii) This mechanism is described as

> Cross through **A**, as each reactant is first order, but this is not a mechanism type.

☐ A first order

☐ B S_N1

☐ C S_N2

☐ D electrophilic substitution **(1 mark)**

The Arrhenius equation

 Maths skills

1 The Arrhenius equation is $k = A_e^{-E_A/RT}$

(a) Identify each of the terms in the equation.

k .....Rate constant

A .....Pre-exponential

E_a .....Activation energy

R .....Gas constant

T .....Temperature........   **(3 marks)**

(b) The logarithmic form of this equation is $\ln k = \ln A - E_a/RT$.

Hydrogen iodide decomposes:

$2HI \rightarrow H_2 + I_2$

In an experiment measuring this decomposition, the following data was collected.

$T/°C$	$k/mol\,dm^3\,s^{-1}$	$\ln k$	T/K	$\frac{1}{T}/K^{-1}$
283	3.52×10^{-7}			
356	3.02×10^{-5}			
393	2.19×10^{-4}			
427	1.16×10^{-3}			
508	3.95×10^{-2}			

(i) Calculate $\ln k$, convert the temperature to K and calculate $\frac{1}{T}$. Add this
data to complete the table.   **(3 marks)**

(ii) Plot $\ln k$ on the y-axis against $\frac{1}{T}$ on the x-axis on
the graph paper below.   **(3 marks)**

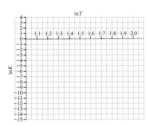

> Draw your graph to use as
> much of the graph paper
> as possible, and then draw
> a large triangle to get an
> accurate figure for the
> gradient.

(iii) Determine the intercept on the y-axis of your graph, and use this and the
logarithmic form of the Arrhenius equation to find A.

Intercept on y-axis ....

....

A = ...   **(3 marks)**

Guided (iv) Determine the gradient of your graph, and use this and the logarithmic
form of the Arrhenius equation to find E_a. $R = 8.31\,J\,K^{-1}\,mol^{-1}$.

Gradient ....

gradient = $-E_a/R$ ....

.... E_a = ... $kJ\,mol^{-1}$ **(4 marks)**

Exam skills 7

1 Nitrogen dioxide decomposes when heated

$$2NO_2(g) \rightarrow 2NO(g) + O_2(g)$$

The concentration of nitrogen dioxide can be measured as time proceeds.

Time/ s	$[NO_2]$/ mol dm^{-3}	Time/ s	$[NO_2]$/ mol dm^{-3}
0	1.00	200	0.48
50	0.79	250	0.43
100	0.65	300	0.38
150	0.55	350	0.34

(a) Plot a graph on the grid of $[NO_2]$ against time.

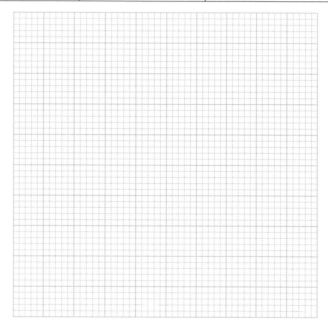

(3 marks)

(b) The concentration–time graph is a curve, showing that the order with respect to NO_2 could be

> If the order were 0, as the concentration of NO_2 reduced the rate would not be affected, and the gradient would be constant.

1   zero order   2   first order   3   second order

☐ A   only 1   ☐ C   only 2 and 3

☐ B   only 2   ☐ D   1, 2 and 3   **(1 mark)**

 2 In the reaction

$$2H_2 + 2NO \rightarrow N_2 + 2H_2O$$

the following rate data was collected.

Experiment	Initial $[H_2]$/ mol dm^{-3}	Initial $[NO]$/mol dm^{-3}	Initial rate/ mol dm^{-3} s^{-1}
1	2.0×10^{-3}	6.0×10^{-3}	6.0×10^{-3}
2	3.0×10^{-3}	6.0×10^{-3}	9.0×10^{-3}
3	6.0×10^{-3}	1.0×10^{-3}	0.5×10^{-3}

Deduce the order with respect to H_2 and to NO.

Experiments 1 and 2: [H2] x 1.5, [NO] x 1, rate x ....

order w.r.t. H_2 ....

order w.r.t. NO ....   **(4 marks)**

Finding the equilibrium constant

Maths skills

1 Hydrogen gas reacts with iodine gas to make hydrogen iodide.

$$H_2 + I_2 \rightleftharpoons 2HI$$

1.00 mol hydrogen and 1.00 mol iodine were mixed in a sealed flask and left until equilibrium was established. At equilibrium, 0.88 mol HI had formed.

(a) Suggest what would be observed as the equilibrium is established.

.... **(1 mark)**

> What colour is iodine gas?

(b) Write an expression for K_c.

.... **(1 mark)**

Guided

(c) Calculate the moles of hydrogen and of iodine at equilibrium.

The moles of hydrogen required to form 0.88 mol HI = ... mol

The moles of hydrogen remaining = 1.00 – .... = ... mol

Moles of iodine = ... mol   **(2 marks)**

(d) Use you answers to part (c) to calculate K_c.

....

....

.... $K_c =$ .... **(2 marks)**

(e) The units of K_c for this equilibrium are

☐   A $mol\,dm^{-3}$

☐   B $mol\,dm^{-3}\,s^{-1}$

☐   C $mol^2\,dm^{-6}$

☐   D   no units   **(1 mark)**

> How many moles are on each side of the equation?

(f) The equilibrium expression involves concentrations in $mol\,dm^{-3}$. Explain why for **this equilibrium** you do **not** need to know the volume of the sealed flask.

....

....

.... **(2 marks)**

(g) At a different temperature, 2.00 mol hydrogen and 1.00 mol iodine were mixed and left, until 0.75 mol hydrogen iodide had formed.

Calculate K_c.

> Follow the steps from parts (b)–(d) but remember that the amounts of each reactant at equilibrium are different.

....

.... $K_c =$ .... **(4 marks)**

Calculating the equilibrium constant, K_c

Guided 1 Describe the difference between a homogeneous and a heterogeneous equilibrium.

In a homogeneous equilibrium, all of the reactants and products are in the

same ...., but in a heterogeneous equilibrium

.... **(2 marks)**

2 Give the expressions for the equilibrium constants for

> Solids and liquids are not included in equilibrium expressions.

(a) $N_2(g) + 3H_2(g) \rightleftharpoons 2NH_3(g)$

....   **(1 mark)**

(b) $CaCO_3(s) \rightleftharpoons CaO(s) + CO_2(g)$

....   **(1 mark)**

(c) For the reaction $2SO_2 + O_2 \rightleftharpoons 2SO_3$, $K_c = 280 \, mol^{-1} \, dm^3$ at 1000 K.

K_c for $2SO_3 \rightleftharpoons 2SO_2 + O_2$ at 1000 K is

☐ A not able to be found from the data provided

☐ B $280 \, mol \, dm^{-3}$

☐ C $280 \, mol^{-1} \, dm^3$

☐ D $0.003\,57 \, mol \, dm^{-3}$ **(1 mark)**

Maths skills

Guided 3 Ethanol reacts with ethanoic acid in the presence of an acid catalyst as follows:

> Some NaOH reacts with the catalyst, but this is a fixed amount and has been taken into account in working out the volume of NaOH required to neutralise the ethanoic acid.

$CH_3COOH(aq) + C_2H_5OH(aq) \rightleftharpoons CH_3COOC_2H_5(aq) + H_2O(l)$

3.0 g ethanoic acid and 2.3 g ethanol were allowed to reach equilibrium. The equilibrium mixture was titrated with $1.0 \, mol \, dm^{-3}$ sodium hydroxide. $16.7 \, cm^3$ of NaOH were required to neutralise the ethanoic acid at equilibrium.

Calculate K_c.

Step 1: mol NaOH = ...

Step 2: mol ethanoic acid at equilibrium = ...

Step 3: mol ethanoic acid at start = ...

Step 4: mol ethanoic acid reacted in forming

> Water is included in K_c, step 4 gives you the mol of ester and the mol of water.

       equilibrium = step 3 – step 2 = ...

Step 5: mol ethanol at start = ...

Step 6: mol ethanol at equilibrium = step 5 – step 4 = ...

Step 7: K_c = ...   **(6 marks)**

Calculating K_p

Maths skills

1 A mixture is made of 0.5 mol nitrogen, 0.1 mol oxygen and 0.2 mol argon.

(a) Calculate the mole fraction of each gas.

> **Maths skills** Give your answer as the simplest fraction or as a decimal.

....

Nitrogen: .... Oxygen: .... Argon: ....   **(2 marks)**

(b) If the total pressure of the mixture is 4 atmospheres, calculate the partial pressure of each gas.

....

Nitrogen: .... Oxygen: .... Argon: ....   **(1 mark)**

Maths skills

2 The substance R decomposes: $R(g) \rightleftharpoons 2S(g)$

A 0.40 mol sample of R is placed in a sealed flask, and left until equilibrium is established. At equilibrium, 0.10 mol of S had formed. The total pressure at equilibrium was 1000 Pa.

Guided

(a) Calculate the moles of R present at equilibrium.

To form 0.10 mol of S, ...mol of R are required.

This leaves 0.40 − .... = ...mol of R at equilibrium.   **(1 mark)**

(b) Calculate the mole fraction of each gas.

....

R: .... S: ....   **(2 marks)**

(c) Calculate the partial pressure of each gas.

....

R: .... S: ....   **(2 marks)**

(d) Write an expression for K_p and calculate its value, including a unit.

....

....

.... K_p = ...   **(3 marks)**

(e) Units for K_p for the equilibrium $NH_4Cl(s) \rightleftharpoons NH_3(g) + HCl(g)$ could be

> What substances are not included in K_p?

☐   A   Pa

☐   B Pa^2

☐   C $mol^{-1} Pa^2$

☐   D $mol\, Pa^{-2}$ **(1 mark)**

The equilibrium constant under different conditions

1 For an equilibrium reaction involving gases, where the forward reaction is endothermic, if the temperature is raised then

1 K_c will increase

2 K_p will increase

3 K_c and K_p are unchanged

Which of the statement(s) are correct?

☐ A only 1

☐ B only 2

☐ C only 1 and 2

☐ D only 3   **(1 mark)**

2 Sulfur dioxide reacts with oxygen to form sulfur trioxide.

$2SO_2(g) + O_2(g) \rightleftharpoons 2SO_3(g)$

At a certain temperature T, $K_c = 40 \, mol^{-1} \, dm^3$

(a) Some sulfur dioxide and oxygen are mixed at temperature T in a sealed container of volume $2.0 \, dm^3$ and left. After a period of time it is found that the moles of each substance are: $SO_2 = 2.0 \, mol$ $O_2 = 0.80 \, mol$ $SO_3 = 7.5 \, mol$

   (i) Write the expression for K_c and calculate the value of the expression using the numbers of moles given.

   ....

   ....

   ....

   ....   **(3 marks)**

⟩Guided⟩ (ii) Explain why the value calculated in part (i) does not match the value given of $40 \, mol^{-1} \, dm^3$.

   The gases have not been left for enough time so ....   **(1 mark)**

(b) In a separate experiment, a sample of sulfur dioxide and oxygen is left in a container until equilibrium is reached.

   Explain, with reference to K_c, in which direction the equilibrium will move if there is a reduction in pressure.

   > As the temperature has not changed, K_c must remain the same.

   ....

   ....

   ....

   ....   **(3 marks)**

Brønsted–Lowry acids and bases

1 (a) What is a Brønsted–Lowry acid?

.... **(1 mark)**

(b) In the reaction

$$HNO_3 + 2H_2SO_4 \rightleftharpoons NO_2^+ + H_3O^+ + 2HSO_4^-$$

which of the reactants is/are acting as a Brønsted–Lowry acid?

☐ A   only HNO_3

☐ B   only H_2SO_4

☐ C HNO_3 and H_2SO_4

☐ D   neither of the reactants

> Which reactant has lost H+ ions in this reaction? It helps to look at the products.

(1 mark)

2 For the equilibrium

$$NH_3 + HBr \rightleftharpoons NH_4^+ + Br^-$$

identify the reactant that is a base and its conjugate acid.

Base: .... Conjugate acid: .... **(2 marks)**

3 (a) Dilute hydrochloric acid is neutralised by sodium hydroxide solution.

    (i) Write the full balanced equation for this reaction, including state symbols.

    .... **(3 marks)**

    (ii) Write the ionic equation for this reaction, including state symbols.

    .... **(2 marks)**

    (iii) Explain what is meant by **spectator ions** and identify all of the spectator ion(s) in this reaction.

> Look at the ions that are unchanged in the ionic equation in part (ii).

    ....

    .... **(2 marks)**

(b) Dilute nitric acid is neutralised by potassium hydroxide solution.

    (i) Write the full balanced equation for this reaction, including state symbols.

    .... **(3 marks)**

    (ii) Write the ionic equation for this reaction, including state symbols.

    .... **(2 marks)**

    (iii) Suggest why the enthalpy change when hydrochloric acid is neutralised by sodium hydroxide solution is the same as that when nitric acid is neutralised by potassium hydroxide solution. The same volume and concentration of each of the four solutions is used.

    .... **(1 mark)**

pH

1 (a) Define pH.

.... **(1 mark)**

(b) Hydrochloric acid is a strong acid.

(i) What is meant by a **strong** acid?

.... **(1 mark)**

Maths skills

(ii) Calculate the pH of $0.01\,mol\,dm^{-3}$ hydrochloric acid.

.... pH = ... **(1 mark)**

Maths skills

Guided

(iii) Calculate the pH of the mixture made when $25.0\,cm^3$ $0.01\,mol\,dm^{-3}$ hydrochloric acid is mixed with $20.0\,cm^3$ $0.005\,mol\,dm^{-3}$ sodium hydroxide solution.

Step 1: equation is .... so a .... ratio

Step 2: mol of HCl = ... mol

Step 3: mol of NaOH = ... mol

Step 4: after mixing, mol of H^+ = ... = ... mol

in .... cm^3 mixture

Step 5: $[H^+]$ = ... = ... $mol\,dm^{-3}$

Step 6: pH = ... = ... **(4 marks)**

2 A solution of a monobasic strong acid has a pH of −1.

(a) What is meant by **monobasic**?

.... **(1 mark)**

(b) The concentration of the acid, in $mol\,dm^{-3}$, is

☐ A   not possible to calculate because a negative pH is not possible

☐ B   −1

☐ C   −log(−1)

☐ D   10 **(1 mark)**

Maths skills

3 $100\,cm^3$ of a solution of a monobasic strong acid has a pH of 1.0.

Calculate the volume of water that would have to be added to raise the pH to 2.0.

| Calculate the mol of H^+ in the original solution, then $[H^+]$ in the diluted solution. |

....

.... volume = ...cm^3 **(4 marks)**

....

....

The ionic product of water

1 (a) Write an equation for the dissociation of pure water.

.... **(1 mark)**

(b) Give the expression for the ionic product of water, K_w.

.... **(1 mark)**

(c) $K_w = 1.0 \times 10^{-14}$ mol^2 dm^{-6} at 25 °C, and $K_w = 4.5 \times 10^{-15}$ mol^2 dm^{-6} at 15 °C.

(i) Calculate [H$^+$] in pure water at 15 °C.

> [H$^+$] = [OH$^-$] in pure water

....

....[H$^+$] = ...mol dm^{-3} **(2 marks)**

(ii) Calculate the pH of pure water at 15 °C.

....pH = ... **(1 mark)**

(iii) The pH of pure water at 25 °C is 7.0.

The pH of pure water at 15 °C is different because

☐ **A** pure water is only neutral at 25 °C

☐ **B** the dissociation of water is endothermic

☐ **C** the dissociation of water is exothermic

☐ **D** water molecules cannot dissociate if the temperature is too low. **(1 mark)**

Guided **2** Calculate the pH of 0.001 mol dm^{-3} sodium hydroxide solution at 15 °C.
$K_w = 4.5 \times 10^{-13}$ mol^2 dm^{-6}

Step 1: [OH$^-$] = ...

Step 2: K_w = ...

Step 3: [H$^+$] = ...

Step 4: pH = ... **(2 marks)**

3 Calculate the pH of the mixture made when 100 cm^3 of 0.015 mol dm^{-3} sodium hydroxide solution is mixed with 75 cm^3 of 0.0080 mol dm^{-3} dilute hydrochloric acid.
$K_w = 1.0 \times 10^{-14}$ mol^2 dm^{-6}

....

....

....

....

....pH = ... **(4 marks)**

The acid dissociation constant

1 The pK_a value for two carboxylic acids at 25°C is given.

Ethanoic	4.75
Chloroethanoic	2.85

(a) (i) Write the equation showing the dissociation of ethanoic acid in water.

.... **(1 mark)**

(ii) Write an expression for K_a of ethanoic acid.

.... **(1 mark)**

Maths skills

(iii) Calculate the value of K_a of ethanoic acid at 25°C.

> **Maths skills** $K_a = 10^{-pKa}$
> Press 10^x then – the value of pK_a.

.... K_a = ... $mol\,dm^{-3}$ **(2 marks)**

(b) (i) Draw the displayed formula of one molecule of chloroethanoic acid.

(1 mark)

Guided

(ii) Is chloroethanoic acid stronger or weaker than ethanoic acid? Suggest why.

Chloroethanoic acid has a .... pK_a value which shows that

....

....

This is because chlorine has a high electronegativity, so ....

.... **(3 marks)**

2 For methanoic acid, at 25°C, $K_a = 1.8 \times 10^{-4}\,mol\,dm^{-3}$.

Maths skills

(a) Calculate, using the usual assumptions, the pH of $0.20\,mol\,dm^{-3}$ methanoic acid at 25°C.

....

....

.... pH = ... **(3 marks)**

(b) Which statement is **not** true about the acid dissociation constant, K_a?

☐ A   Water can be omitted from K_a because it is in large excess.

☐ B K_a does not apply if one of its salts is dissolved in the acid.

☐ C   For a given acid, only a change of temperature can alter K_a.

☐ D   The higher the pK_a value the weaker the acid. **(1 mark)**

Approximations made in weak acid pH calculations

 Maths skills

1 Phenol, C_6H_5OH, is a weak, monobasic acid. It reacts with sodium hydroxide to form sodium phenoxide. At 25°C, $K_a = 1.0 \times 10^{-10}\,mol\,dm^{-3}$.

(a) Write the equation for the neutralisation of phenol with sodium hydroxide.

....   **(1 mark)**

(b) Calculate the pH of $0.25\,mol\,dm^{-3}$ phenol at 25°C.

....

....

.... pH = ...   **(3 marks)**

(c) State clearly two approximations that you have made in this calculation.

....

....

....   **(2 marks)**

(d) Trichloroethanoic acid, at 25°C, has $K_a = 0.22\,mol\,dm^{-3}$.

The calculation of the pH of $0.25\,mol\,dm^{-3}$ trichloroethanoic acid using the standard approximations gives a value pH = 0.63.

Will the value you calculated in part (b) of the pH of phenol or the pH value of trichloroethanoic acid, pH = 0.63, be more accurate? Explain, in terms of one of the approximations, why.

> Phenol is very weak, but tricholoroethanoic acid is stronger.

....

....

....

....   **(3 marks)**

2 A $0.40\,mol\,dm^{-3}$ solution of a weak monobasic acid, HX, has a pH of 3.20.

(a) K_a for HX is

☐ A $10^{-3.20} \times 0.40$

☐ B $\dfrac{-\log(3.20)}{0.40}$

☐ C $\dfrac{0.40}{10^{-3.20}}$

☐ D $\dfrac{(10^{-3.20})^2}{0.40}$ **(1 mark)**

(b) What is the name given, in the acid dissociation equilibrium, to the X^- ion?

....   **(1 mark)**

Buffers

1 (a) What is a buffer?

....

....

.... **(2 marks)**

(b) Which mixture could make a buffer solution?

> A weak acid and its salt must be present after any reaction has occurred.

☐ A   hydrochloric acid and sodium hydroxide

☐ B   ethanoic acid and sodium hydroxide

☐ C   hydrochloric acid and sodium ethanoate

☐ D   sodium hydroxide and sodium ethanoate   **(1 mark)**

2 The blood is buffered.

(a) The normal pH range of blood is

☐ A   5.6–5.7

☐ B   6.0–6.1

☐ C   6.95–7.05

☐ D   7.35–7.45   **(1 mark)**

(b) Explain why it is important that blood is buffered.

....

....

.... **(2 marks)**

(c) Identify the substance that buffers blood that is

(i)   a weak acid: ....   **(1 mark)**

(ii) a conjugate base ion: ....   **(1 mark)**

>Guided> 3 A buffer can be made by mixing propanoic acid with potassium propanoate solution.

(a) Write equations showing the dissociation of these two substances in water.

$CH_3CH_2COOH(aq) + H_2O(l) \rightleftharpoons$ ....

$CH_3CH_2COOK(aq) \rightarrow$ ....   **(2 marks)**

(b) Explain how this buffer is effective when some dilute acid is added to the mixture.

....

....

....

.... **(2 marks)**

Buffer calculations

>Guided>

1 A buffer solution is made by mixing 150 cm³ of 0.100 mol dm⁻³ propanoic acid with 75 cm³ 0.150 mol dm⁻³ sodium propanoate solution. For propanoic acid, $K_a = 1.3 \times 10^{-5}$ mol dm⁻³.

Calculate the pH of the buffer.

Stage I: mol propanoic acid = ... = ... mol

Stage 2: [CH₃CH₂COOH] in mixture = ... = ... mol dm⁻³

Stage 3: mol sodium propanoate = ... = ... mol

Stage 4: [CH₃CH₂COONa] in mixture = ... = ... mol dm⁻³

Stage 5: K_a = ...

Stage 6: [H⁺] = ... = ... mol dm⁻³

Stage 7: pH = ...   **(4 marks)**

2 100 cm³ of 0.500 mol dm⁻³ propanoic acid is mixed with 100 cm³ 0.500 mol dm⁻³ sodium propanoate solution. For propanoic acid, $K_a = 1.3 \times 10^{-5}$ mol dm⁻³.

The pH of the buffer mixture is

☐ A 1.3×10^{-5}

☐ B $-\log \sqrt{(1.3 \times 10^{-5} \times 0.500)}$

☐ C $\log \left(\dfrac{100}{1000} \times 0.500 \right)$

☐ D $-\log (1.3 \times 10^{-5})$ **(1 mark)**

3 A buffer solution is made by mixing 250 cm³ of 0.200 mol dm⁻³ ethanoic acid with 150 cm³ 0.100 mol dm⁻³ sodium ethanoate solution. For ethanoic acid, $K_a = 1.8 \times 10^{-5}$ mol dm⁻³.

(a) Calculate the pH of the buffer.

....

....

.... pH = ...   **(4 marks)**

(b) 50 cm³ of 0.100 mol dm⁻³ sodium hydroxide solution is added to this buffer mixture.

Calculate the pH after the addition of the sodium hydroxide solution.

....

....

....

.... pH = ...   **(4 marks)**

> The sodium hydroxide reacts with the propanoic acid, reducing the mol of acid; and sodium propanoate is formed in the reaction, increasing the mol of sodium propanoate.

pH titration curves

1 The diagram shows the pH titration curve when $1\,cm^3$ portions of $0.100\,mol\,dm^{-3}$ sodium hydroxide solution are added to $20.0\,cm^3$ of $0.150\,mol\,dm^3$ ethanoic acid.

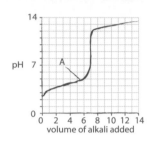

(a) The neutralisation volume of sodium hydroxide solution is

☐ A $20\,cm^3$

☐ B $25\,cm^3$

☐ C $30\,cm^3$

☐ D $35\,cm^3$ **(1 mark)**

(b) Use the diagram to find the pH at the neutralisation point.

....   **(1 mark)**

⟩**Guided**⟩   (c) Explain why the pH changes only gradually in the part of the graph labelled **A**.

In this part, the sodium hydroxide solution will have neutralised some of

the ethanoic acid, so the substances present are ....

....   **(2 marks)**

2 Sketch below the pH curve obtained by adding $50\,cm^3$ of $1\,mol\,dm^{-3}$ hydrochloric acid (a strong acid) from a burette to $25\,cm^3$ $1\,mol\,dm^{-3}$ ammonia solution (a weak base). Label your axes and give scales. Include values on the sketch of the start and end pH and the neutralisation volume.

(6 marks)

Indicators

1 The pH of a solution can be measured using an indicator or a pH meter.

(a) Explain how a pH meter is calibrated.

....

.... **(2 marks)**

(b) The pH range of some indicators is given:

Indicator	pH range
methyl yellow	2.9–4.0
methyl orange	3.1–4.4
cresol red	7.2–8.8
phenolphthalein	8.3–10
indigo carmine	11.4–13.0

A pH titration curve is given to the right.

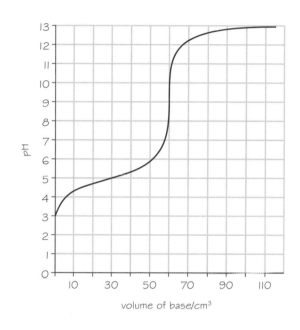

volume of base/cm³

Guided

(i) Identify **all** of the indicator(s), if any, which would be suitable for this titration, explaining how you make the choice.

Indicator(s): ....

Explanation: The vertical part of the pH graph ....

.... **(2 marks)**

(ii) Explain why **none** of the indicators is suitable for a weak acid–weak base titration.

> What is the shape of the pH curve at pH = 7?

....

.... **(2 marks)**

2 Two diagrams are given of the main species present in a solution of phenolphthalein at pH −1 and pH 5. Suggest the colour of the species present at each pH.

pH = −1 colour = ...	pH = 5 colour = ...
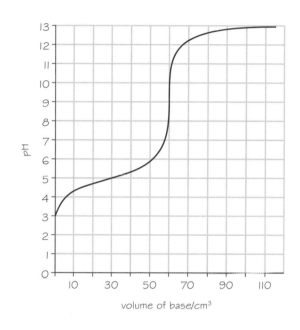	

(2 marks) 97

Exam skills 8

1 In the production of sulfuric acid, one of the processes involves the following reaction

$2SO_2(g) + O_2(g) \rightleftharpoons 2SO_3(g)$ $\Delta H = -198\,kJ\,mol^{-1}$

(a) Explain the effect on the yield of sulfur trioxide of

   (i) increasing the pressure

   ....

   .... **(2 marks)**

   (ii) increasing the temperature.

   ....

   .... **(2 marks)**

Maths skills

(b) 2 mol of sulfur trioxide is placed in a container of volume 150 dm³.

At equilibrium 1.45 mol remain. Calculate K_p, in kPa⁻¹, if the total pressure is 35 kPa.

....

....

....

.... $K_p =$ .... kPa⁻¹   **(4 marks)**

(c) Sulfur trioxide is changed into sulfuric acid, H_2SO_4.

   (i) Sulfuric acid is a dibasic acid.

   What is meant by **dibasic**?

   .... **(1 mark)**

Maths skills

   (ii) Calculate the pH of a mixture of 25.0 cm³ 0.100 mol dm⁻³ sulfuric acid and 75.0 cm³ of 0.100 mol dm⁻³ sodium hydroxide solution.

   > Calculate the moles of each substance and then the excess of NaOH moles.

   ....

   .... **(3 marks)**

Maths skills

   (iii) Sketch the pH curve you would get by adding 50.0 cm³ of 0.100 mol dm⁻³ sodium hydroxide solution to 25.0 cm³ 0.100 mol dm⁻³ hydrochloric acid.

   (3 marks)

The Born–Haber cycle

1 The diagram below shows a Born–Haber cycle for the formation of lithium fluoride, with some relevant data, all in $kJ\,mol^{-1}$.

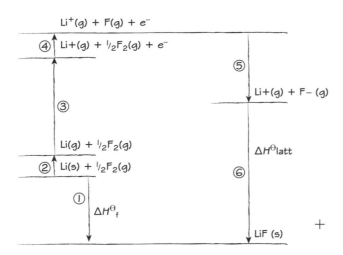

Enthalpy change of atomisation of lithium / $kJ\,mol^{-1}$	159
Enthalpy change of atomisation of fluorine/ $kJ\,mol^{-1}$	79
First ionisation energy of lithium/ $kJ\,mol^{-1}$	520
First electron affinity of fluorine/ $kJ\,mol^{-1}$	−328
Enthalpy of formation of lithium fluoride/ $kJ\,mol^{-1}$	−616

(a) Step 4 is

> How many moles of fluorine atoms are made? How many moles of F–F bonds are broken?

    1   the enthalpy change of atomisation of fluorine

    2   the molar bond enthalpy of F–F

    3   the enthalpy of vaporisation of fluorine

Which of the statement(s) are correct?

☐   A   only 1

☐   B   only 2

☐   C   only 1 and 2

☐   D   1, 2 and 3   **(1 mark)**

(b) Give the names for:

    (i)   Step 1:..   **(1 mark)**

    (ii)   Step 3:...   **(1 mark)**

    (iii) Step 5:..   **(1 mark)**

Maths skills

Guided

(c) Use the data to calculate the lattice enthalpy for lithium fluoride.

lattice enthalpy = step 6 = – step 5 – ....

....

lattice enthalpy = ...$kJ\,mol^{-1}$...............   **(2 marks)**

Factors affecting lattice enthalpy

1 Data is given in the table of the lattice enthalpy of some ionic compounds, all in $kJ\,mol^{-1}$.

	F⁻	Cl⁻	Br⁻	I⁻
Li⁺	−1036	−853	−807	−757
Na⁺	−923	−787	−747	−704
K⁺	−821	−715	−682	−649
Rb⁺	−785	−689	−660	−630

(a) Give the definition of the lattice enthalpy of lithium fluoride.

....

....

.... **(2 marks)**

Guided

(b) State and explain the trend in the lattice enthalpies from lithium fluoride to lithium iodide.

Moving down Group 7, the size of the halide ions gets

....

.... **(3 marks)**

> Use unambiguous language – use 'less exothermic' or 'more exothermic' as appropriate.

(c) State and explain the trend in the value of the lattice enthalpies from lithium fluoride to rubidium fluoride.

....

....

....

.... **(3 marks)**

(d) (i) Give the electronic configurations of a sodium ion and of a magnesium ion.

Na⁺ ....

Mg²⁺ .... **(2 marks)**

(ii) Explain how the lattice enthalpy of magnesium oxide would compare with that of sodium bromide.

> Compare the size of and charges on the ions.

....

....

....

.... **(3 marks)**

The enthalpy change of solution

1 Data about calcium chloride and its ions is given below.

Lattice enthalpy of calcium chloride/ kJ mol^{-1}	-2255
Enthalpy change of hydration of calcium ions/ kJ mol^{-1}	-1650
Enthalpy change of hydration of chloride ions/ kJ mol^{-1}	-338

Maths skills

(a) Write equations for the changes whose enthalpy change is the

   (i)   lattice enthalpy of calcium chloride

   .... **(2 marks)**

   (ii) enthalpy change of hydration of calcium ions

   .... **(2 marks)**

   (iii) enthalpy change of solution of calcium chloride

   .... **(2 marks)**

(b) (i) Draw a Hess' law cycle for dissolving calcium chloride, using the changes in (a) and the enthalpy change of hydration of chloride ions.

(2 marks)

   (ii) Use the data provided to calculate the enthalpy change of solution of calcium chloride.

                                         Use your Hess' law cycle.

   ....

   ....$\Delta_{sol}H =$ ....kJ mol^{-1} **(2 marks)**

(b) Comparing the enthalpy change of hydration of sodium ions with calcium ions,

   1   that of calcium is more exothermic because the ion is more highly charged

   2   that of calcium is more exothermic because the ion is larger

   3   only that of sodium is exothermic because all sodium compounds are soluble.

Which of the statement(s) are correct?

☐ A   only 1

☐ B   only 2

☐ C   only 1 and 2

☐ D   only 3   **(1 mark)**

Entropy

1 Which of these reactions lead/s to a decrease in entropy of the system?

> Consider the states of the reactants and products.

1   $2NaHCO_3(s) + H_2SO_4(aq) \rightarrow Na_2SO_4(aq) + H_2O(l) + CO_2(g)$

2   $AgNO_3(aq) + NaCl(aq) \rightarrow NaNO_3(aq) + AgCl(s)$

3   $Mg(s) + 2HNO_3(aq) \rightarrow Mg(NO_3)_2(aq) + H_2(g)$

☐ A   only 1

☐ B   only 2

☐ C   only 1 and 3

☐ D   only 3

2 Iron(III) oxide is reduced in the blast furnace by carbon monoxide.

Data about the relevant substances is given in the table.

Substance	$\Delta_f H^\ominus$ / kJ mol^{-1}	$S^\ominus$ / J K^{-1} mol^{-1}
Fe_2O_3	-824	87.4
CO	-111	197.6
Fe	0	27.3
CO_2	-394	213.6

(a) Write the balanced equation for the reaction.

.... **(1 mark)**

>Guided>

(b) Calculate the standard enthalpy change of reaction.

$\Delta_r H^\ominus = \Delta_f H^\ominus$ (products) $- \Delta_f H^\ominus$ (reactants) $= $ ....

....

....

....$\Delta_r H^\ominus = $ .... kJ mol^{-1} **(3 marks)**

(c) Calculate the standard entropy change of reaction.

....

....

....$\Delta S^\ominus = $ .... J K^{-1} mol^{-1} **(3 marks)**

(d) Explain the **sign** of the entropy change that you have calculated.

....

.... **(2 marks)**

(e) The liquid iron can be cooled and solidified. Explain, in terms of iron atoms, why this leads to a decrease in entropy.

.... **(1 mark)**

Free energy

1 (a) Give the equation relating free energy, enthalpy and entropy.

.... **(1 mark)**

(b) (i) What is meant by the term **feasible**?

....

.... **(1 mark)**

(ii) For a reaction to be feasible at **all** temperatures, the values of the enthalpy change and of the entropy change must be

		Enthalpy change	Entropy change
☐	A	negative	negative
☐	B	negative	positive
☐	C	positive	negative
☐	D	positive	positive

(1 mark)

Maths skills

2 Hydrogen is produced by reacting methane with steam, $CH_4(g) + H_2O(g) \rightleftharpoons CO(g) + 3H_2(g)$

Data about these substances is given below.

Substance	$\Delta_f H^\ominus$ / kJ mol^{-1}	$S^\ominus$/ J K^{-1} mol^{-1}
$CH_4(g)$	−74.9	186
$H_2O(g)$	−241.8	189
$CO(g)$	−110.5	198
$H_2(g)$	0	131

(a) Calculate the standard enthalpy change of reaction.

....

....

.... $\Delta_r H^\ominus$ = ... kJ mol^{-1} **(2 marks)**

(b) Calculate the standard entropy change of reaction.

....

.... $\Delta S^\ominus$ = ... J K^{-1} mol^{-1} **(1 mark)**

(c) Calculate the free energy change of reaction at 125 °C.

....

.... ΔG = ... kJ mol^{-1} **(2 marks)**

> Remember to convert temperature to K and ΔS to kJ.

Guided

(d) Calculate the temperature above which this reaction becomes feasible.

$\Delta G = 0$ when ΔH = ...

....

.... T = ... K   **(2 marks)**

Redox

1 Peroxodisulfate ions, $S_2O_8^{2-}$, react with bromine.

$$5S_2O_8^{2-} + Br_2 + 6H_2O \rightarrow 2BrO_3^- + 12H^+ + 10SO_4^{2-}$$

(a) Calculate the oxidation number of

(i)   Br in Br_2 ....

(ii) Br in BrO_3^- .... **(2 marks)**

(b) Identify the species that has been reduced in the reaction, and the reducing agent.

Reduced species: ....

Reducing agent: .... **(2 marks)**

2 Cr^{3+} ions can be oxidised to CrO_4^{2-} ions in alkaline solution.

(a) What is meant by **oxidation**?

....

.... **(1 mark)**

>Guided> (b) Construct a half-equation for this reaction.

Cr^{3+} + ... OH^- → **(2 marks)**

(c) Another chromium ion, $Cr_2O_7^{2-}$ is used as an oxidising agent in acidic conditions. Its half-equation is

$$Cr_2O_7^{2-} + 14H^+ + 6e^- \rightarrow 2Cr^{3+} + 7H_2O$$

(i)   Calculate the oxidation number of chromium in

$Cr_2O_7^{2-}$ .... Cr^{3+} .... **(2 marks)**

(ii) What colour change is seen in this reaction?

| You have met this reaction in organic chemistry. |

.... **(1 mark)**

(iii) Ethanol is oxidised by $Cr_2O_7^{2-}$ ions.

Construct the half-equation for the oxidation of ethanol, in the presence of water, to form ethanoic acid and H^+ ions.

.... **(2 marks)**

(iv) Combine the equation given in part (c) and your answer to part (iii) to produce an equation for the oxidation of ethanol by dichromate ions.

| Cancel down H^+ ions and H_2O molecules that appear on both sides of the equation. |

....

....

.... **(2 marks)**

Redox titrations

1 'Iron tablets', a dietary supplement, contain iron(II) sulfate. A method to determine the percentage by mass of iron in the tablets is given.

> **Step 1:** Weigh accurately ten iron tablets, and place in a conical flask containing 75 cm³ of approximately 1.50 mol dm⁻³ sulfuric acid. Shake and leave the tablets to dissolve.
>
> **Step 2:** Filter the solution into a volumetric flask and make up to 100 cm³.
>
> **Step 3:** Pipette 25.0 cm³ of the solution into a conical flask and add 20.0 cm³ of the dilute sulfuric acid with a measuring cylinder.
>
> **Step 4:** Titrate with 0.0200 mol dm⁻³ potassium manganate(VII) solution.
>
> **Step 5:** Calculate the concentration of the iron(II) ions in the solution and hence the percentage by mass of iron in the tablets.

(a) Suggest why the tablets have an outer coating that is insoluble in water but breaks down slowly in acid.

> How are the tablets taken?

....

.... **(1 mark)**

(b) Why is it not necessary to know the exact concentration of the sulfuric acid?

.... **(1 mark)**

>**Guided** (c) Suggest how step 2 can be carried out to get the most accurate results.

The residue should be ....

.... **(1 mark)**

(d) What would be the effect on the **titre** value (volume of potassium manganate(VII) solution needed) if 21.0 cm³ of sulfuric acid was added in step 3?

☐ A   The titre value would be too high by an unknown amount.

☐ B   The titre value would be too low by an unknown amount.

☐ C   The titre value would be unchanged.

☐ D   The titre value would be too high by 5%. **(1 mark)**

(e) Explain why an indicator has not been added in the method.

.... **(1 mark)**

(f) If the tablets' mass was 1.96 g and mean titre value 24.25 cm³, calculate the percentage by mass of iron in the tablets. The products of the redox reaction are Fe^{3+} and Mn^{2+}.

> Follow these steps:
> 1. moles of MnO_4^-
> 2. moles of Fe^{2+} in 25 cm³
> 3. moles of Fe^{2+} in 100 cm³
> 4. mass of Fe^{2+}
> 5. % by mass

....

....

....

.... **(5 marks)**

Electrochemical cells

1 The diagram below shows a standard hydrogen electrode.

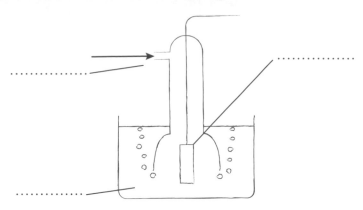

(a) Complete the labels on the diagram.   **(3 marks)**

(b) Give the standard conditions required for the standard hydrogen electrode.

> 3 points are needed here to get the 3 marks.

....

....

....   **(3 marks)**

Guided (c) Describe how the standard hydrogen electrode is used with a half-cell to measure the standard electrode potential of the half-cell.

The SHE is connected to ....

....

....   **(2 marks)**

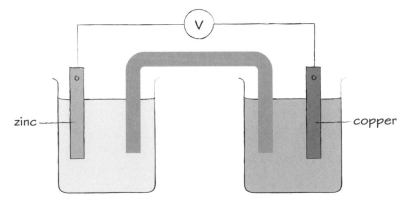

2 The diagram shows a zinc–copper cell.

In this cell, state

(a) a suitable solution to use in the zinc half-cell ....   **(2 marks)**

(b) a suitable salt bridge ....   **(1 mark)**

(c) the purpose of the salt bridge ....   **(1 mark)**

(d) the direction of the electron flow ....   **(1 mark)**

(e) the electrode at which reduction occurs ....   **(1 mark)**

Measuring and using standard electrode potentials

Use these values to answer the questions.

$$
\begin{array}{lll}
 & & E^{\ominus}/V \\
Cl_2(g) + 2e^- & \rightarrow \quad 2Cl^-(aq) & +1.36 \\
NO_3^-(aq) + 3H^+(aq) + 2e^- & \rightarrow \quad HNO_2(aq) + H_2O(l) & +0.94 \\
Fe^{3+}(aq) + e^- & \rightarrow \quad Fe^{2+}(aq) & +0.77 \\
I_2(aq) + 2e^- & \rightarrow \quad 2I^-(aq) & +0.54 \\
Fe^{2+}(aq) + 2e^- & \rightarrow \quad Fe(s) & -0.44
\end{array}
$$

1 (a) Describe how the NO_3^-, H^+ half-cell would be set up under standard conditions.

....

....

.... **(3 marks)**

(b) Which species listed below is the most powerful oxidising agent?

> The oxidising agent is reduced, so gains electrons.

☐ A Cl_2 ☐ C Fe^{2+}

☐ B Cl^- ☐ D Fe^{3+} **(1 mark)**

(c) The $E^{\ominus}$ value in volts when iron is oxidised **to Fe^{2+}** by chlorine is

☐ A $+1.36 - 0.44$ ☐ C $+1.36 - 0.88$

☐ B $+1.36 + 0.44$ ☐ D $+1.36 + 0.88$ **(1 mark)**

2 (a) Write the ionic equation for the reaction between Fe^{3+} ions and I^- ions, and for the reaction between Fe^{2+} ions and I^- ions, and calculate the $E^{\ominus}$ values for each of these reactions.

....

....

....

....

....

.... **(6 marks)**

> **Guided**

(b) If Fe^{3+} ions are mixed with iodide ions, explain the final iron species that will be formed.

The answer to part (a) shows that the positive $E^{\ominus}$ ....

....

.... **(2 marks)**

Predicting feasibility

Use these values to answer the questions.

1 Some potassium manganate(VII) solution is mixed with some hydrogen peroxide solution.

$$E^{\ominus}/V$$

$V^{3+}(aq) + e^-$	$\rightarrow$	$V^{2+}(aq)$	-0.26
$SO_4^{2-}(aq) + 4H^+(aq) + 2e^-$	$\rightarrow$	$H_2SO3(aq) + H_2O(l)$	$+0.17$
$VO^{2+}(aq) + 2H^+(aq) + e^-$	$\rightarrow$	$V^{3+}(aq) + H_2O(l)$	$+0.34$
$O_2(g) + 2H^+(aq) + 2e^-$	$\rightarrow$	$H_2O_2(aq)$	$+0.68$
$Fe^{3+}(aq) + e^-$	$\rightarrow$	$Fe^{2+}(aq)$	$+0.77$
$VO_2^+(aq) + 2H^+(aq) + e^-$	$\rightarrow$	$VO^{2+}(aq) + H_2O(l)$	$+1.00$
$2IO_3^-(aq) + I2H^+(aq) + IOe^-$	$\rightarrow$	$I_2(aq) + 6H_2O(l)$	$+1.19$
$MnO_4^-(aq) + 8H^+(aq) + 5e^-$	$\rightarrow$	$Mn^{2+}(aq) + 4H_2O(l)$	$+1.52$

(a) Use the half-equations above to construct an ionic equation for this reaction.

....

.... **(1 mark)**

(b) Calculate the $E^{\ominus}$ value for this reaction.

.... $E^{\ominus}$ = ... V   **(1 mark)**

(c) Explain whether or not this reaction is feasible.

.... **(2 marks)**

(d) What observations would be made if some potassium manganate(VII) solution is mixed with an excess of hydrogen peroxide solution.

....   **(2 marks)**

> The ionic equation, and your use of potassium manganate(VII) in titrations should help.

2 The $E^{\ominus}$ value for a reaction involving gases and solutions has been calculated and it has been found to be feasible. However, no reaction is observed.

An **incorrect** possible explanation is that

☐ A the reaction is very slow   ☐ C the activation energy is low

☐ B the solutions are not $1.00\,mol\,dm^{-3}$ ☐ D the pressure is not $10^5\,Pa$ **(1 mark)**

3 Some oxygen gas is bubbled through an acidified solution of V^{2+} ions.

Explain which vanadium species will be present when no further reaction occurs.

Write ionic equations, and calculate $E^{\ominus}$ values, for each reaction that occurs.

....

....

....

.... **(5 marks)**

Storage and fuel cells

1　A diagram of a storage cell is shown.

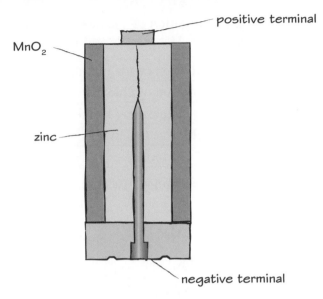

positive terminal

MnO_2

zinc

negative terminal

(a)　Identify which part of the cell is the anode and which is the cathode.

Anode: .... cathode: ....　**(2 marks)**

(b)　The relevant half-reactions for such a cell are

$ZnO + H_2O + 2e^- \rightarrow Zn + 2OH^-$　　　$E^\ominus = -1.28\,V$

$2MnO_2 + H_2O + 2e^- \rightarrow Mn_2O_3 + 2OH^-$　$E^\ominus = +0.15\,V$

(i)　Write the equation for the reaction that occurs in the cell.

....

.... **(1 mark)**

> Remember to reverse one half-reaction so that overall $E^\ominus$ is positive.

(ii)　Calculate $E^\ominus$ for the reaction occurring.

.... $E^\ominus = $ .... V　**(1 mark)**

(iii)　A voltmeter is attached to a cell identical to that above.

Suggest why the reading is not equal to the value calculated in part (ii).

....　**(1 mark)**

 2　(a)　How does a fuel cell differ from a storage cell?

The fuel in a fuel cell is ....

The voltage from a fuel cell is ....　**(2 marks)**

(b)　The half-equations for one fuel cell are:

$4H_2O + 4e^- \rightarrow 2H_2 + 4OH^-$　　$E^\ominus = -0.83\,V$

$2H_2O + O_2 + 4e^- \rightarrow 4OH^-$　　$E^\ominus = +0.40\,V$

Write the overall equation and calculate the $E^\ominus$ value.

....

.... $E^\ominus = $ .... V　**(2 marks)**

Exam skills 9

1 The enthalpy change of formation of the theoretical compound MgCl can be determined by using a Born–Haber cycle and the data below, all in $kJ\,mol^{-1}$.

Enthalpy change of atomisation of magnesium / $kJ\,mol^{-1}$	146
Enthalpy change of atomisation of chlorine / $kJ\,mol^{-1}$	122
First ionisation energy of magnesium / $kJ\,mol^{-1}$	738
First electron affinity of chlorine / $kJ\,mol^{-1}$	−349
Lattice enthalpy of MgCl / $kJ\,mol^{-1}$	−753

(a) The equation for the reaction whose enthalpy change is the enthalpy of formation of MgCl is

☐ A $2Mg(s) + Cl_2(g) \rightarrow 2MgCl(s)$

☐ B $Mg(s) + Cl(g) \rightarrow MgCl(s)$

☐ C $Mg^+(g) + Cl^-(g) \rightarrow MgCl(s)$

☐ D $Mg(s) + \frac{1}{2}Cl_2(g) \rightarrow MgCl(s)$ **(1 mark)**

(b) Draw a Born–Haber cycle and calculate the enthalpy of formation of MgCl.

....

.... $\Delta_f H =$ .... $kJ\,mol^{-1}$ **(5 marks)**

⟩**Guided**⟩   (c) Suggest why, when magnesium is heated in chlorine, **only** $MgCl_2$ is formed.

The enthalpy of formation of $MgCl_2$ is ....

....   **(1 mark)**

(d) Magnesium chloride can be electrolysed to form chlorine gas.

Outline how the standard electrode potential of $Cl_2(g) + 2e^- \rightarrow 2Cl^-(aq)$ can be measured.

....

....

....   **(5 marks)**

The transition elements

1 (a) Define the following terms.

(i) d-block element

.... **(1 mark)**

(ii) transition element

.... **(1 mark)**

(b) Use your periodic table to list the **symbols** of **all** the elements in **Period 4** that are

(i) transition elements

.... **(1 mark)**

(ii) d-block elements but **not** transition elements

.... **(1 mark)**

(iii) **not** d-block elements.

.... **(1 mark)**

> These are the s- and p-block elements.

Guided **2** (a) Give the electrons-in-boxes electronic configuration for

(i) cadmium

	4d					5s
[Kr]						

(1 mark)

(ii) cadmium's only ion, Cd^{2+}

	4d					5s
[Kr]						

(1 mark)

(b) Explain whether or not cadmium is a transition element.

.... **(1 mark)**

3 Which of the following electronic configurations is correct?

			3d					4s
☐	A	Ti	[Ar] ↑ ↑ ↑ ↑ ☐					↑↓
☐	B	V	[Ar] ↑↓ ↑ ☐ ☐ ☐					↑↓
☐	C	Cr^+	[Ar] ↑ ↑ ↑ ↑ ☐					↑
☐	D	Mn^{2+}	[Ar] ↑ ↑ ↑ ↑ ↑					☐

(1 mark)

Properties of transition elements

1 Transition elements and their ions can act as catalysts.

   (a) What is a catalyst?

   ....

   .... **(3 marks)**

   (b) Transition elements and their ions can act as catalysts because

   1   they are very unreactive, so they do not get used up in reactions

   2   they readily undergo redox reactions

   3   they have variable oxidation states.

   Which of the statement(s) are correct?

   ☐ A   only 1

   ☐ B   only 1 and 2

   ☐ C   only 2 and 3

   ☐ D   only 3   **(1 mark)**

   (c) Hydrogen peroxide decomposes as follows:

   $2H_2O_2(aq) \rightarrow 2H_2O(l) + O_2(g)$

   A catalyst for this reaction is the black solid manganese(IV) oxide, MnO_2. Design an experiment that shows that manganese(IV) oxide is a catalyst for the reaction.

   > **Practical skills** You need to show that the reaction is faster, and that the catalyst has not been used up.

   ....

   ....

   ....

   .... **(4 marks)**

2 A solution is made in a small beaker of a transition metal compound, A. The pale green crystals of A dissolve to make a green solution, B. When dilute sodium hydroxide solution is added to B, a green precipitate C forms. After leaving the beaker to stand, a brown colouration D is seen at the surface of the mixture in the beaker.

   (a) Identify, by name or formula,

   > Only one transition element has compounds that can be green or brown.

   the transition **element** found in A: ....

   The transition element **ion** found in solution B:

   ....

   Precipitate C: ....

   Substance D: ....   **(4 marks)**

   (b) What type of reaction has occurred in the formation of D?

   .... **(1 mark)**

Complex ions

1 The diagram shows the ion $[Cr(NH_3)_6]^{3+}$.

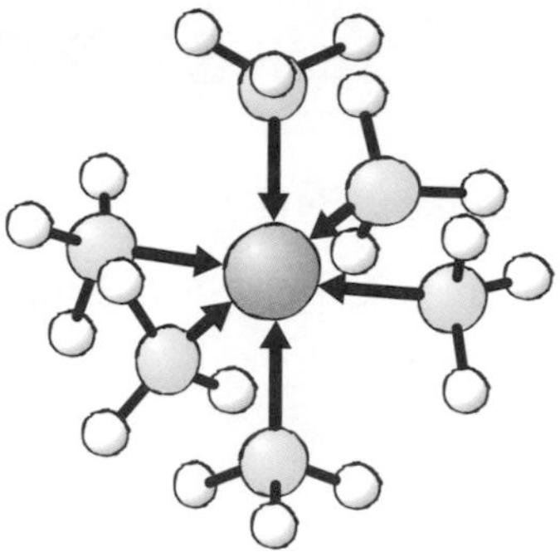

(a) On the diagram:

(i) clearly label a covalent bond that is **not** a coordinate bond   **(1 mark)**

(ii) clearly label a coordinate bond   **(1 mark)**

(iii) circle **one** ligand.   **(1 mark)**

(b) In this complex, the ligand is

☐ A   monodentate

☐ B   bidentate

☐ C   tridentate

☐ D   multidentate.   **(1 mark)**

(c) State and explain the coordination number of this complex.

....   **(2 marks)**

(d) State the shape of this complex.

....   **(1 mark)**

⟩**Guided**⟩ 2 The ligand 'en' is $NH_2CH_2CH_2NH_2$.

(a) Explain how this molecule is able to form bonds with a transition metal ion.

The nitrogen atoms on the molecule have ....

....

....   **(2 marks)**

(b) The complex ion $[Ni(en)_3]^{2+}$

1   has a coordination number of 6

2   has a trigonal planar shape

3   has an empirical formula $NiCNH_4$

Which of the statement(s) are correct?

☐ A   only 1

☐ B   only 2

☐ C   only 1 and 2

☐ D   only 1 and 3   **(1 mark)**

4-fold coordination and isomerism

1 Platin has the formula $Pt(NH_3)_2Cl_2$.

(a) Draw a diagram of *cis*-platin and *trans*-platin below.

cis   trans

(3 marks)

(b) State the meaning of the term **cis–trans isomerism** of a transition metal complex.

....

.... **(2 marks)**

(c) Name the shape that these two isomers exhibit.

.... **(1 mark)**

(d) (i) Give an example of a complex ion that has a coordination number of four and has a different shape from platin.

.... **(1 mark)**

> Remember the shape formed by molecules with four bonds.

(ii) Draw a diagram of this complex ion, and name the shape.

Shape .... **(2 marks)**

2 Isomerism can exist in complex ions.

1 $[Ni(en)_3]^{2+}$ exists as optical isomers.

2 The ion $CuCl_4^{2-}$ exhibits stereoisomerism.

3 the isomers *cis*-platin and *trans*-platin are used as anti-cancer drugs.

Which of the statement(s) are correct?

☐ A only 1

☐ B only 2

☐ C only 1 and 2

☐ D only 1 and 3 **(1 mark)**

Precipitation reactions

Practical skills

1 A compound of manganese(II) is dissolved in water to make solution **A**. The solution is divided into two portions in test tubes. Drops of sodium hydroxide solution are added to each portion so that precipitate **B** forms. One test tube is then left to stand. The other test tube has an excess of sodium hydroxide solution added.

Guided

(a) What is a precipitate?

When two solutions are mixed, ...

.... **(1 mark)**

(b) (i) State the colour of solution **A**.

.... **(1 mark)**

    (ii) Give the formula of the coloured ion in **A**.

| This is an aqua-ion. |

.... **(1 mark)**

(c) (i) State the colour of precipitate **B**.

.... **(1 mark)**

    (ii) Write the ionic equation for the formation of **B**.

.... **(2 marks)**

(d) The test tube darkens when left to stand.

This is because

| What could the open mixture have reacted with? |

☐ A   an acid–base reaction has occurred

☐ B   a ligand exchange reaction has occurred

☐ C   a redox reaction has occurred

☐ D   the original reaction was slow and has now finished. **(1 mark)**

(e) For the test tube that has excess sodium hydroxide added, describe what would be observed, if anything.

.... **(1 mark)**

Practical skills

2 Describe fully the observations that you would make when ammonia solution is added dropwise and then in excess to a solution of a copper(II) compound, giving ionic equations.

....

....

....

....

.... **(6 marks)**

Ligand substitution reactions

1　When ammonia solution is added to a solution of a chromium(III) compound, reactions occur in two stages.

(a)　In the first stage, as ammonia is added dropwise a reaction occurs.

(i)　Describe the observation that you would make.

....　**(1 mark)**

(ii)　Write the ionic equation for this reaction, giving full formulae for the complex ions.

....　**(3 marks)**

(iii)　What type of reaction has occurred?

☐　A　acid–base

☐　B　ligand exchange

☐　C　redox reaction

☐　D　decomposition　**(1 mark)**

(b)　In the second stage, as ammonia is added in excess, a further reaction occurs.

(i)　Describe the observation that you would make.

....　**(1 mark)**

(ii)　Write the ionic equation for this reaction, giving full formulae for the complex ions.

....　**(3 marks)**

(iii)　What type of reaction has occurred?

☐　A　acid–base

☐　B　ligand exchange

☐　C　redox reaction

☐　D　neutralisation　**(1 mark)**

⟩Guided⟩　2　(a)　Describe how haemoglobin in the blood enables transport of oxygen to cells.

Oxygen molecules bind to ....

....

....　**(2 marks)**

(b)　Explain why the build-up of carbon monoxide in a room is dangerous.

> CO is involved in a ligand exchange reaction.

....

....　**(3 marks)**

Redox reactions of transition elements

1   A transition metal compound, **A**, consists of blue crystals. Some of these crystals are dissolved in water to make a blue solution, **B**. A solution containing iodide ions is added dropwise until no further change occurs. The result is a liquid with a brown appearance, **C**, containing a white precipitate, **D**.

(a) What transition metal is found in A?

....   **(1 mark)**

(b) Give the formula of the ion with the blue colour in **B.**

....   **(1 mark)**

(c) Which molecule gives the brown appearance to C?

....   **(1 mark)**

(d) Give the formula of the precipitate D.   The precipitate is an iodide.

....   **(1 mark)**

(e) What type of reaction has occurred?

☐ A   acid–base

☐ B   ligand exchange

☐ C   redox reaction

☐ D   neutralisation   **(1 mark)**

2   Some $K_2Cr_2O_7$ crystals are dissolved in water. To the solution a colourless liquid alcohol is added. The mixture is warmed and a colour change occurs.

(a) (i)   Calculate the oxidation state of chromium in $K_2Cr_2O_7$.

....   **(1 mark)**

(ii) Using your answer to part (i), give the IUPAC name for $K_2Cr_2O_7$.

....   **(1 mark)**

(b) What colour change is seen?

from .... to ....   **(2 marks)**

(c) (i)   Explain, in terms of redox, what has occurred in   You may be more familiar with this
this reaction, and what may be deduced about   reaction from organic chemistry.
the alcohol.

....

....

....   **(3 marks)**

Exam skills 10

1   Nickel, copper and zinc are adjacent elements in the d-block.

(a) (i)   Give the electronic configuration of an atom of each element.

Ni ....   Zn ....

Cu ....   **(3 marks)**

(ii)   Most copper compounds are blue or green, whilst zinc compounds are white.

State why.

....   **(1 mark)**

(b)   Ions of transition elements may undergo acid–base, ligand exchange and redox reactions.

You are given a solution of copper(II) sulfate, divided into three portions.

Describe a reaction that you could do with this solution that involves each of these three types of reaction. You should give the reagents and colour changes, but equations are not required.

Acid–base: ....   **(2 marks)**

Ligand exchange: ....   **(2 marks)**

Redox: ....   **(2 marks)**

(c) (i)   Nickel, in oxidation state 0, forms a complex with four carbon monoxide molecules.

Draw, and give the name of, two possible shapes of the complex.

....   **(4 marks)**

(ii)   An aqueous solution containing nickel(II) ions forms a complex with four cyanide ions, CN^-, as the only ligands.

Write an ionic equation for this reaction.

....   **(2 marks)**

(iii) Hydroxocobalamin, a molecule related to vitamin B_{12}, contains cobalt, and is used as an antidote to cyanide poisoning. Cyanide ions form complexes with both iron and cobalt.

Cyanide ions behave similarly to carbon monoxide.

Suggest why cyanide ions are toxic and how hydroxocobalamin works as an antidote.

....

....

....   **(3 marks)**

The bonding in benzene rings

1 The scientist Kekulé proposed a structure for the molecule benzene with three C=C and three C–C bonds.

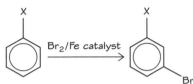

(a) Give the molecular and the empirical formula of a benzene molecule.

Molecular formula: .... **(1 mark)**

Empirical formula: .... **(1 mark)**

Guided (b) Describe how a σ and a π bond are formed.

In a σ bond, the pair of ....

....

....

.... **(2 marks)**

(c) (i) State and explain the observation you would expect when reacting bromine with benzene, if benzene had a Kekulé structure, naming the expected product and the type of reaction.

....

....

.... **(4 marks)**

(ii) State the observation you make when bromine is shaken with benzene.

.... **(1 mark)**

(iii) Explain the reason for the difference in the observation in parts (i) and (ii).

....

....

.... **(2 marks)**

How are the π bond electrons arranged in benzene?

(d) (i) Write the equation for the complete hydrogenation of benzene.

.... **(1 mark)**

(ii) Name the product formed in part (i).

.... **(1 mark)**

Reactions of benzene rings

1 Benzene reacts with bromine to form bromobenzene.

   (a) What is the electrophile in this reaction?

   ☐ A Br_2

   ☐ B Br

   ☐ C Br^+

   ☐ D Br^- **(1 mark)**

   (b) Write an equation

       (i)   showing the formation of the electrophile.

       ....   **(2 marks)**

       (ii) showing the overall equation for the reaction.

       ....   **(2 marks)**

   (c) If two hydrogen atoms are substituted, a molecule with the molecular formula
        $C_6H_4Br_2$ will form. Draw and name three isomers with this formula.

        > Each molecule is based
        > on a benzene ring where
        > the relative positions of
        > the two bromine atoms
        > are different.

        ....

        ....   **(4 marks)**

   (d) Consider the molecule below.

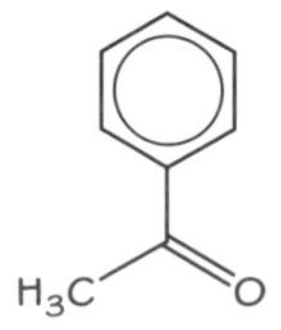

        Name the two substances that when mixed with benzene would lead to the
        formation of this molecule.

        ....   **(2 marks)**

Electrophilic substitution reactions

1   Benzene will react with 2-chloropropane and aluminium chloride in an electrophilic substitution reaction.

(a) What is the electrophile in this reaction?

☐ A $CH_3CHClCH_3$

☐ C $CH_3CHClCH_3^+$

☐ B $AlCl_3$

☐ D $CH_3CH^+CH_3$ **(1 mark)**

(b) (i) Write an equation showing the formation of the electrophile.

....   **(1 mark)**

(ii) State and explain the role of the aluminium chloride.

....

....   **(2 marks)**

(c) Draw the mechanism for the reaction.

(3 marks)

2   A molecule is shown that is formed by the alkylation of benzene.

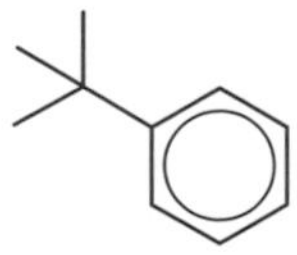

Draw and name the chloroalkane that would react with benzene to form the molecule.

....   **(2 marks)**

Comparing the reactivity of alkenes and aromatic compounds

1 (a) In an experiment, samples of hex-2-ene, benzene or cyclohexene are placed in three boiling tubes. To each, some bromine is added and the mixture is shaken. Complete the table below showing the structure of these molecules, the observation when the mixtures are shaken and the structure of the product of any reaction.

Molecule	Structure	Observation on shaking with bromine	Structure of product, if any, formed
hex-2-ene			
benzene			
cyclohexene			

(5 marks)

(b) Explain the difference in the reactivity of hex-2-ene and benzene in the reaction above.

The electron density in π bonds is delocalised over the ring ....

....

....

.... (3 marks)

2 Which of the below could be formed from an unsaturated hydrocarbon and bromine and no other reactant?

☐ A $CH_3CH_2BrCH_2Br$

☐ B $CH_3CHBrCHBrCHBrCH_2Br$

☐ C C_6H_5Br

☐ D (structure showing cyclohexane ring with Br substituent)

> Try removing the bromine and seeing if an alkene is left.

(1 mark)

Phenol

1   Three reactions, **A**, **B** and **C** are shown.

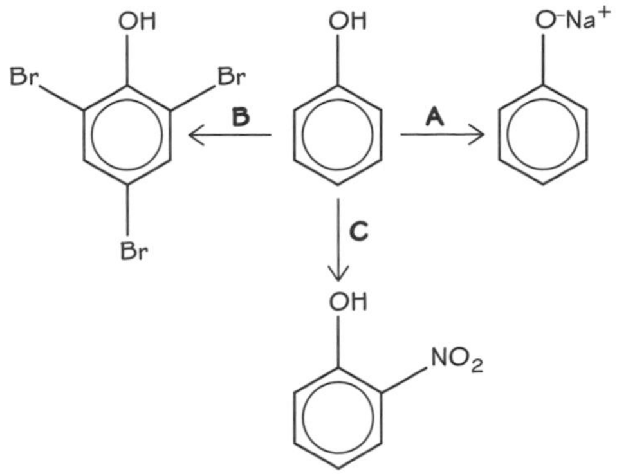

(a) (i) Write a balanced equation for reaction **A**.

....   **(1 mark)**

(ii) What type of reaction is this?

....   **(1 mark)**

(b) Draw and name the isomer of the product shown in **C** most likely also to be formed in the reaction.

(2 marks)

(c) (i) What observation is made when reaction **B** is carried out?

....   **(2 marks)**

> Remember to give the starting colour of the bromine solution.

(ii) What type of reaction has occurred?

☐ A neutralisation

☐ B substitution

☐ C addition

☐ D elimination   **(1 mark)**

> Guided

(iii) State and explain the different reaction conditions required when bromine reacts with benzene and with phenol.

When bromine reacts with benzene a catalyst ....

....

....   **(3 marks)**

Directing effects in benzene

1 The molecule shown below is reacted with bromine and an iron catalyst. When a single hydrogen atom is substituted by a bromine atom, more than one product forms. The major product is shown.

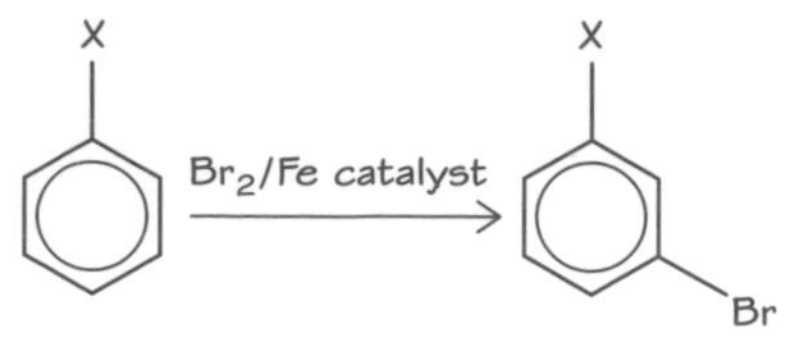

(a) What is X?

☐ A   H

☐ B   OH

☐ C NO_2

☐ D NH_2 **(1 mark)**

(b) What is the non-organic product of the reaction?

....   **(1 mark)**

2 Phenol and benzene both react with nitric acid.

(a) (i) State the conditions required for benzene to react with nitric acid.

....   **(2 marks)**

> Remember to mention **both** the acid and any other required reagent.

(ii) State the **difference** in the conditions required for phenol, rather than benzene, to react with nitric acid.

....

> Comment on each of the conditions you mentioned in part (i).

....   **(2 marks)**

(iii) Explain why different conditions must be used in parts (i) and (ii).

....

....   **(3 marks)**

(iv) A molecule has the formula C_7H_7Br, and forms a cream precipitate when warmed with silver nitrate solution. Draw the structure of the molecule.

(1 mark)

Exam skills 11

1 Consider the molecules below.

P	Q	R	S
	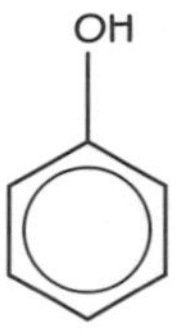 OH	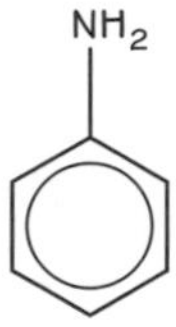 NH$_2$	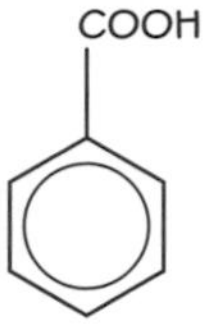 COOH

(a) The molecule(s) that can neutralise sodium hydroxide are

☐ A   Q, R and S

☐ B   Q and S

☐ C   R

☐ D   none of them   **(1 mark)**

(b) **Q** and **R** both react with bromine.

(i) Describe what you see when this reaction occurs.

....

.... **(2 marks)**

> Give the colours seen before and after the reaction.

(ii) Name the type of reaction that occurs and the product that forms when **R** reacts with an excess of bromine.

Type of reaction: ....   **(1 mark)**

Name of product: ....   **(1 mark)**

(iii) **R** and **S** do not react readily to form a substituted amide. Suggest why.

....

....

.... **(2 marks)**

(iv) **R** and **S** can be made to form an amide using a catalyst. Draw the displayed formula of the amide that would form.

> Think about the link that forms when a carboxylic acid reacts with an amine.

                                                                                          (2 marks)

Aldehydes and ketones

1 You are given samples of the following molecules:

A	B	C	D

(a) Give the observations you would make when

   (i) warming each substance, with acidified potassium dichromate solution.

A: ....

B: ....

C: ....

D: ....   **(4 marks)**

   (ii) warming each substance, with 2,4 dinitrophenylhydrazine.

A: ....

B: ....

C: ....

D: ....   **(4 marks)**

(b) Molecule **B** gives a silver mirror when warmed with Tollens' reagent.

   (i) What change occurs in the oxidation number of silver particles in the reaction?

   ☐ A   +1 to 0

   ☐ B   +2 to 0

   ☐ C   +1 to +2

   ☐ D   There is no change   **(1 mark)**

> **Guided**

   (ii) Write an equation, using [O], showing the change occurring to **B** in this reaction.

   .... + ... [O] → .... **(1 mark)**

> [O] is used to show oxygen atoms originating from the oxidising agent.

(c) Molecule **C** reacts with sodium cyanide in acidic solution.

   Draw and name the organic product.

   .... **(2 marks)**

Nucleophilic addition reactions

1 Which of the following species **cannot** act as a nucleophile?

☐   A H^-

☐   B CN^-

☐   C Br_2

☐   D H_2O **(1 mark)**

2 Butan-2-one can be converted to butan-2-ol.

(a) Give the formula of the reagent required.

....   **(1 mark)**

(b) In addition to nucleophilic addition, what type of reaction is this?

....   **(1 mark)**

(c) Write the equation for this reaction.

....   **(1 mark)**

3 Consider the mechanism below for the reaction of ethanal with cyanide ions,

> Three arrows are required.

(a) Complete the mechanism by adding the curly arrows.   **(1 mark)**

(b) Give the name of the product.

....   **(1 mark)**

⟩**Guided**⟩ (c) Explain why the product will be a racemic mixture.

The atoms around the carbonyl group are in a ....

....

....

....   **(2 marks)**

(d) Suggest why neither pure HCN(g) nor pure HCN(aq) is used to provide the cyanide ions.

....

....

....   **(2 marks)**

Carboxylic acids

Guided 1 An organic molecule contains carbon, hydrogen and oxygen only. When a sample is added to sodium hydrogencarbonate solution, effervescence is seen.

(a) Explain the observation with the sodium hydrogencarbonate solution.

The effervescence is .... showing that the molecule

contains the functional group .... **(2 marks)**

Maths skills (b) The molecule is found to have 34.6% carbon and 3.85% hydrogen. Calculate the empirical formula of the molecule.

....

....

....

.... **(4 marks)**

(c) The relative molecular mass of the molecule is 104. State the molecular formula, and draw and name the molecule which contains **one** type of functional group only.

.... **(3 marks)**

2 Which of the following observations are correct for a carboxylic acid?

1 When calcium carbonate is added, effervescence is seen.

2 When warmed with Tollens' reagent a silver precipitate forms.

3 When magnesium powder is added, effervescence is seen.

☐ A   only 1 and 2

☐ B   only 1 and 3

☐ C   only 2 and 3

☐ D   1, 2 and 3 **(1 mark)**

3 You are provided with samples of three colourless liquids; an alkene, a carboxylic acid and an aldehyde. Describe test tube reactions to distinguish the substances.

....

....

....

....

....

.... **(6 marks)**

Esters

1   The molecule below is a food additive.

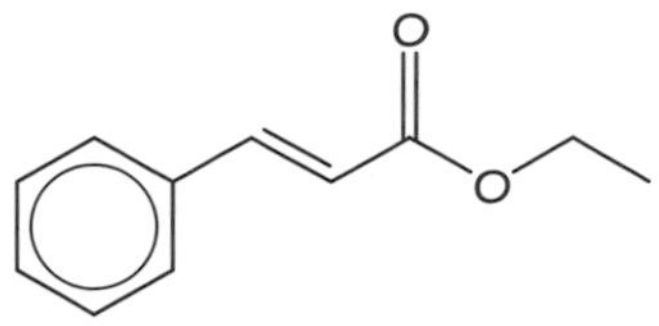

(a)   The name of this molecule is

☐ A   3-benzene-ethylpropenoate

☐ B   3-phenylpropenylethanoate

☐ C   ethyl 1-benzenepropanoate

☐ D   ethyl 3-phenylpropenoate.   **(1 mark)**

> $C_6H_5CH_2CH_2COOH$ is called phenylpropanoic acid.

(b)   Suggest two properties of this substance that would make it suitable as a food additive.

....

....   **(2 marks)**

(c)   The molecule is shaken with bromine. How many bromine atoms would be added to each molecule of ester?

☐ A   1

☐ B   2

☐ C   7

☐ D   8   **(1 mark)**

(d)   (i)   What is hydrolysis?

....

....   **(2 marks)**

(ii)   Describe one way in which an ester can be hydrolysed.

....

....   **(2 marks)**

(iii) Name the alcohol and draw the acid that would be formed by the hydrolysis of the molecule above.

....   **(2 marks)**

Had a go ☐ Nearly there ☐ Nailed it! ☐

Acyl chlorides

1 The compound propanoyl chloride can be formed from a carboxylic acid and sulfurous dichloride.

(a) Draw the structure of propanoyl chloride.

(1 mark)

(b) Give the formula of sulfurous dichloride.

.... **(1 mark)**

(c) (i) Write the equation for the reaction.

Two small molecules are formed in addition to the acyl chloride.

.... **(2 marks)**

(ii) Apart from safety glasses and gloves, explain a safety precaution that you should use if carrying out this synthesis.

.... **(2 marks)**

2 A molecule is shown below.

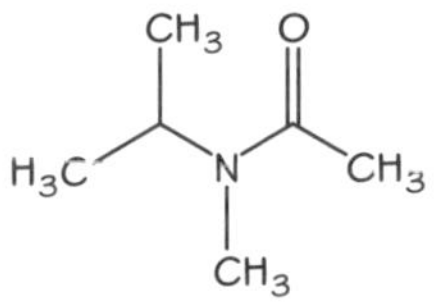

(a) Name the functional group in the molecule.

.... **(1 mark)**

(b) This molecule can be made from an acyl chloride and one other molecule. Draw the acyl chloride and the other molecule.

(2 marks)

(c) An acyl chloride reacts with

☐ A water to form an ester

☐ B an alcohol to form an ester

☐ C ammonia to form an amine

☐ D an amine to form a nitrile. **(1 mark)**

Amines

1 Consider the four molecules below.

A	B	C	D
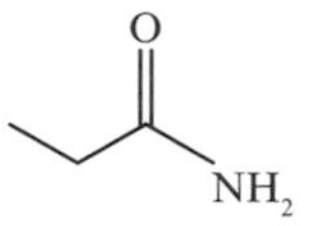	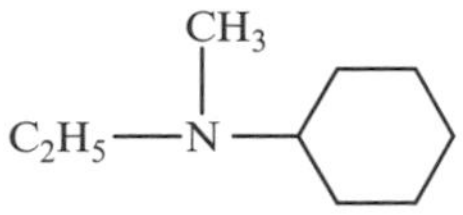	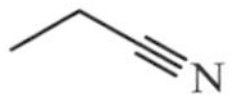	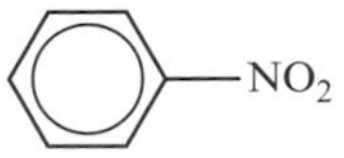

(a) Identify the functional group in each molecule.

A: .... **(1 mark)**

B: .... **(1 mark)**

C: .... **(1 mark)**

D: .... **(1 mark)**

〉**Guided**〉 (b) Is molecule **B** a primary, secondary or tertiary example of its functional group? Explain your answer.

The number of groups joined to ....

....

.... **(2 marks)**

2 A straight-chain molecule **P** has the molecular formula C_3H_9N. It reacts with bromomethane to form **Q**, $C_4H_{11}N$, and some other nitrogen containing organic products, and an acidic molecule **R** which can react with **P** to form **S**, $C_3H_{10}NBr$.

> Consider which elements are found in **P** and the fact that it reacts with **R**.

(a) What is the functional group in **P**?

☐ A amide

☐ B amine

☐ C nitrile

☐ D amino acid **(1 mark)**

(b) Give the structural formulae of **P**, **Q**, **R** and **S**.

P: .... **(1 mark)**

Q: .... **(1 mark)**

R: .... **(1 mark)**

S: .... **(1 mark)**

Amino acids

1 The molecule below is the amino acid ornithine, which is involved in the
   urea cycle in the body.

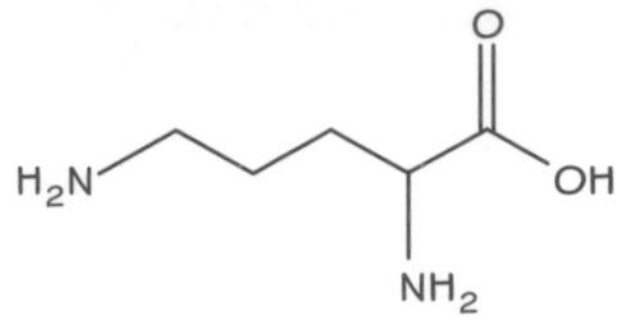

(a) Draw the structure that would be formed if ornithine was added to

   (i)   excess hydrochloric acid.

   (1 mark)

   (ii) sodium hydroxide.

   (1 mark)

   (iii) Ornithine reacts with another amino acid, aspartic acid, to form a molecule
        used as a treatment for liver disease. The molecular formula of aspartic acid is
        $C_4H_7O_4N$. How many structural isomers of $C_4H_7O_4N$ exist that are amino acids?

   ☐ A 1

   ☐ B 2

   ☐ C 3

   ☐ D 4   **(1 mark)**

(b) The IUPAC name of ornithine is

   ☐ A 2,5-diaminopentanoic acid

   ☐ B 1,4-diaminopentanoic acid

   ☐ C pentanoic acid-2-5-diamine

   ☐ D 5-carboxy-hexyl-1,4-diamine.   **(1 mark)**

(c) Decarboxylation, a decomposition reaction, is where a carboxylic acid group
    releases carbon dioxide. Suggest an equation for the decarboxylation of
    ornithine, and name the organic product.

   .... **(2 marks)**

Optical isomers

1   The molecule **X** has the molecular formula $C_4H_7O_3Cl$. When sodium carbonate solution is added to **X**, effervescence is seen. When **X** is warmed with acidified potassium dichromate solution, there is a colour change from orange to green.

(a) (i)   What gas causes the effervescence?

    ....   **(1 mark)**

    (ii)   What can be deduced from the reaction with sodium carbonate?

    ....   **(1 mark)**

(b) (i)   What type of reaction is occurring with acidified potassium dichromate solution?

    ....   **(1 mark)**

    (ii)   Give all of the functional groups that could be in **X** that would cause a colour change with acidified potassium dichromate solution.

    ....   **(2 marks)**

(c) Draw a structure of **X** that

    (i)   exists as optical isomers (draw both isomers) (**P** and **Q**)

                                                                                            (2 marks)

    (ii)   does not exist as an optical isomer (**R**)

                                                                                            (1 mark)

(d) (i)   What is a racemic mixture?

    ....

    ....   **(1 mark)**

    (ii)   Is it possible to distinguish between a pure sample of **R** and a racemic mixture of **P** and **Q** using plane-polarised light? Explain your answer.

    ....

    ....

    ....   **(2 marks)**

Condensation polymers

1 Give a similarity and a difference between addition polymerisation and condensation polymerisation.

....

....

.... **(2 marks)**

2 A diagram of part of a molecule of the polymer Kevlar™ is shown. Kevlar™ has many uses due to its high strength to weight ratio. Kevlar™ is synthesised from a diamine and a diacylchloride.

(a) What type of polymerisation forms Kevlar?

....   **(1 mark)**

(b) Draw the two monomers required to form Kevlar™, giving the name for the diamine only.

diamine	diacyl chloride
....	

(3 marks)

(c) What other product is formed when Kevlar is synthesised?

.... **(1 mark)**

3 The molecule 2-hydroxypropanoic acid can be polymerised.

Draw a molecule of the monomer and the repeat unit of the polymer, and give the type of polymerisation.

> Think how the two functional groups in the monomer would react with each other.

....   **(3 marks)**

Exam skills 12

1 The formula of a polymer is given. This polymer was formed from two monomers.

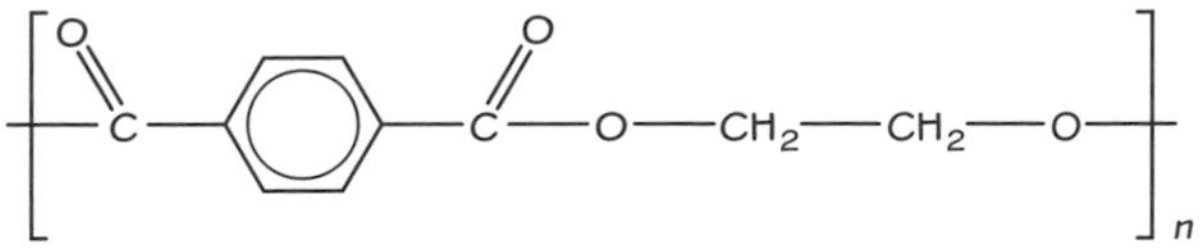

(a) The empirical formula of the polymer is

☐ A $C_{10}H_8O_4$

☐ B $C_{10}H_{10}O_4$

☐ C $C_{10n}H_{8n}O_{4n}$

☐ D $C_5H_4O_2$.   **(1 mark)**

(b) (i) What type of polymer is this?

....   **(1 mark)**

(ii) A small molecule that could not be formed when this
polymerisation occurs is

> What could react to form the functional group in the polymer?

☐ A ammonia

☐ B hydrogen chloride

☐ C water

☐ D poly(ethene).   **(1 mark)**

(c) If this polymer is hydrolysed under alkaline conditions, using KOH, two
products are formed.

(i) Draw the structural formula and name each product.

Molecule 1:   Molecule 2:

> Remember that the monomer formed by hydrolysis might react with alkali.

....   **(4 marks)**

(ii) State the difference in the products formed when the polymer is
hydrolysed under acidic conditions.

....   **(1 mark)**

Carbon–carbon bond formation

1 (a) A primary amine contains 69.0% carbon and 16.1% nitrogen. Calculate the
empirical formula of the amine.

....

....

.... **(3 marks)**

(b) The relative formula mass of the molecule is 87, and the molecule exists as
optical isomers. Draw the structural formula of one molecule, indicating the
chiral carbon with an asterisk.

(2 marks)

(c) This primary amine can be synthesised from a bromoalkane containing
four carbon atoms in two steps. In the first step a molecule with five carbons
is formed.

(i) The functional group of the product of step 1 is an

| How can a four-carbon molecule form the product required? |

☐ A amine

☐ B alcohol

☐ C amide

☐ D nitrile **(1 mark)**

(ii) Give the reagents required to react with the bromoalkane in step 1.

.... **(2 marks)**

(iii) Write the equation for step 2.

.... **(2 marks)**

(iv) Give the catalyst required for step 2.

....

| This metal is often used as a catalyst with the gas involved in step 2. |

(1 mark)

(v) What type of reaction has occurred in each
of these two steps?

	Step 1	Step 2
☐ A	substitution	neutralisation
☐ B	substitution	reduction
☐ C	elimination	substitution
☐ D	addition	reduction

(1 mark)

Purifying organic solids

1 Practical instructions for the recrystallisation of impure benzoic acid are given.

> Step 1: Place a weighed amount of impure benzoic acid crystals into a flask and slowly add boiling water until all the crystals dissolve. **Do not add more water than necessary.**
>
> Step 2: Filter the hot mixture, then allow the filtrate to cool slowly.
>
> Step 3: Collect the solid crystals by vacuum filtration. Wash the crystals with cold water.
>
> Step 4: Transfer the product to a watch glass and allow to dry.
>
> Step 5: Weigh your recrystallised benzoic acid.
>
> Step 6: Measure the melting point of the crystals.

(a) What property of benzoic acid does this purification method depend upon?

Benzoic acid is more .... in hot water .... **(1 mark)**

(b) Why must the minimum amount of water be used in step 1?

....

....

.... **(2 marks)**

(c) Why are the crystals washed in step 3?

.... **(1 mark)**

(d) The procedure is carried out, and in step 6 it is found that the melting point is not sharp, with a different range from the melting point for benzoic acid in a data book. Suggest an error in one of the steps that could cause this result.

> What would cause this effect on the melting point? What step incompletely carried out could lead to this?

....

....

.... **(2 marks)**

(e) The benzoic acid was made from 5.00 g benzaldehyde, and the final mass of benzoic acid, once the error mentioned in part (d) was corrected, was 5.35 g. Calculate the percentage yield.

....

....

.... **(4 marks)**

Had a go ☐ Nearly there ☐ Nailed it! ☐

Predicting the properties and reactions of organic compounds

1 The molecule to the right exists as two enantiomers. One enantiomer smells of spearmint and the other of caraway seeds.

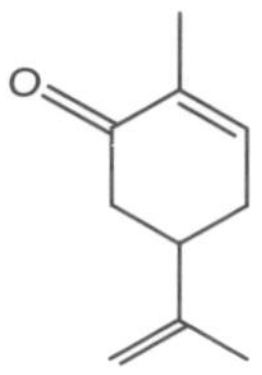

(a) Identify two functional groups found in the molecule.

.... **(2 marks)**

(b) State what you would see when the molecule is

(i) shaken with bromine solution

.... **(1 mark)**

(ii) warmed with acidified potassium dichromate(VI) solution

.... **(1 mark)**

(iii) added to 2,4-dinitrophenylhydrazine

.... **(1 mark)**

(c) How many moles of hydrogen, with a nickel catalyst, will react with one mole of the molecule?

Which groups will be reduced?

☐ A 0

☐ B 1

☐ C 2

☐ D 3 **(1 mark)**

(d) (i) In the figure above identify with an asterisk the chiral carbon. **(1 mark)**

(ii) Suggest why the two molecules have different smells.

A molecule causing an odour binds to a smell receptor in the nose.

....

....

.... **(2 marks)**

(e) Draw the structure of the product formed when the molecule is reacted with steam in the presence of phosphoric acid.

(2 marks)

Summary of organic reactions

1   The table summarises reactants, products, reagents and conditions for some reactions.

Reaction	Reactant	Product(s)	Reagent	Conditions
1	propane	**A1** and **A2**	air	excess oxygen
2	but-2-ene	2,3-dibromobutane	**B**	
3	propan-1-ol	propanal	$K_2Cr_2O_7$ and **C**	**D**
4	propanenitrile	**E**	hydrogen	**F**
5	nitrobenzene	phenylamine	**G**	**H**
6	**I**	pentan-3-ol	$NaBH_4$	
7	**J**	butanoyl chloride	**K**	

(a)   Name the reactant, product(s) or reagent, or specify the condition

A1: .... A2: ....   **(2 marks)**

B: ....   **(1 mark)**

C: ....   **(1 mark)**

D: ....   **(1 mark)**

E: ....   **(1 mark)**

F: ....   **(1 mark)**

G: ....   **(1 mark)**

H: ....   **(1 mark)**

I: ....   **(1 mark)**

J: ....   **(1 mark)**

K: ....   **(1 mark)**

(b)   Explain how condition **D** can be altered to give a different product, naming the organic product.

....

....

....

....   **(3 marks)**

(c)   Which of reactions 1–7 are redox reactions?

.... | Which reagents are oxidising or reducing agents? |

    (2 marks)

Had a go ☐   Nearly there ☐   Nailed it! ☐

Organic synthesis

1 Benzene, C_6H_6, can be converted to 4-chlorophenylamine in three stages.

The starting material, intermediates and product are shown.

A	B
Cl (chlorobenzene structure)	(benzene structure)
C	D
Cl, NH₂ (4-chlorophenylamine structure)	Cl, O₂N (structure)

(a) Arrange the molecules **A**, **B**, **C** and **D** in the order in which they are found in the synthesis.

.... **(1 mark)**

(b) For each of the steps, name the intermediate, giving its letter, and name the necessary reagents/conditions. If a step forms more than one isomer, only the isomer necessary for the next step should be considered.

> Think about which of the options can be formed from benzene.

Step 1:

from benzene (...............) to .... (...............)

....

.... **(3 marks)**

Step 2:

from .... (...............) to .... (...............)

....

.... **(3 marks)**

Step 3:

from .... (...............) to 4-chlorophenylamine (...............)

....

.... **(3 marks)**

Thin layer chromatography

1 (a) Uses of TLC do **not** include

☐ A   monitoring the progress of a reaction

☐ B   finding the number of components in a mixture

☐ C   verifying the identity of a compound

☐ D   distinguishing between two enantiomers.   **(1 mark)**

(b) The diagram shows TLC plates taken from a reaction in which one reactant
(R) forms one product (P). Plate 1 shows a sample of pure R and pure P.
Plates 2, 3 and 4 show samples from the reaction mixture. The samples for
plates 2, 3 and 4 are taken as time passes, in that order.

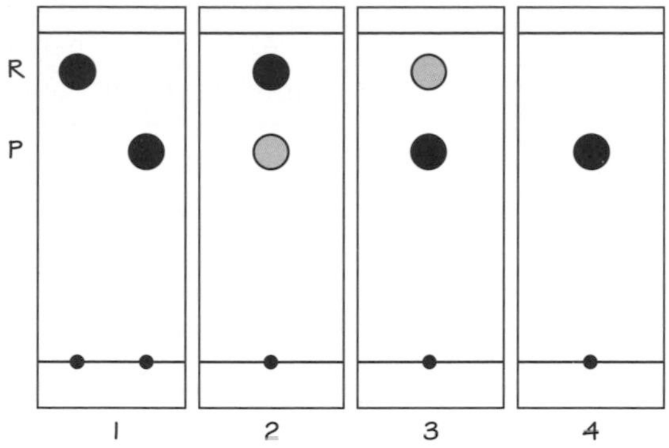

Guided

(i)   Explain the patterns found in these plates.

As time passes, the spot showing R ....

....

.... **(2 marks)**

(ii) What is the purpose of plate 1?

....

.... **(1 mark)**

(iii) Calculate the R_f values of R and P.

....

.... **(3 marks)**

2 Two technicians carry out TLC on the same pure compound and measure the
R_f value. Which of the following is **not** required for the R_f value to be the same?

☐ A   The same solvent

☐ B   The same material for the stationary phase

☐ C   The same thickness of the material making the stationary phase

☐ D   The TLC plate is left in the solvent for the same time   **(1 mark)**

Gas chromatography

1 (a) The diagram shows the results of gas chromatography on a mixture.

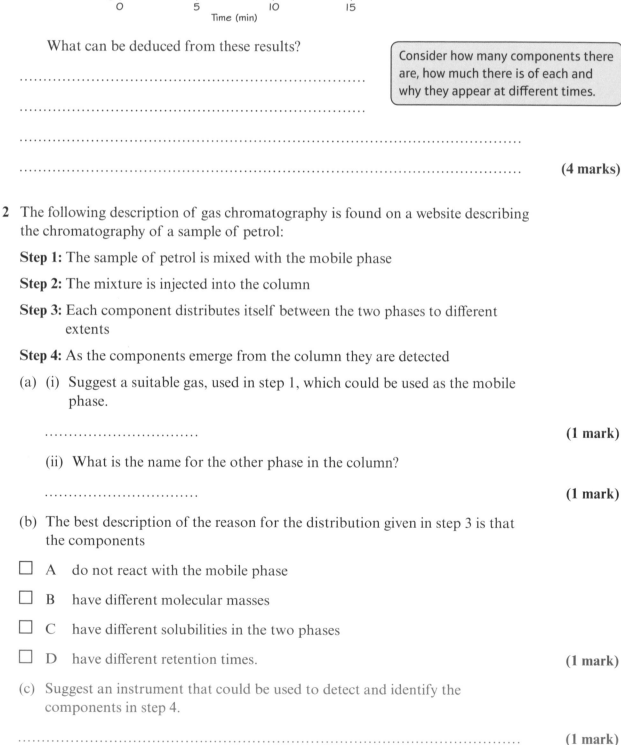

What can be deduced from these results?

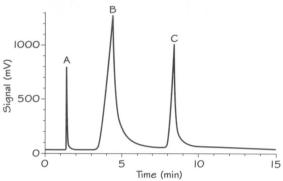
Consider how many components there
are, how much there is of each and
why they appear at different times.

....

....

....

....   **(4 marks)**

2 The following description of gas chromatography is found on a website describing
the chromatography of a sample of petrol:

Step 1: The sample of petrol is mixed with the mobile phase

Step 2: The mixture is injected into the column

Step 3: Each component distributes itself between the two phases to different
extents

Step 4: As the components emerge from the column they are detected

(a) (i) Suggest a suitable gas, used in step 1, which could be used as the mobile
phase.

....   **(1 mark)**

(ii) What is the name for the other phase in the column?

....   **(1 mark)**

(b) The best description of the reason for the distribution given in step 3 is that
the components

☐ A   do not react with the mobile phase

☐ B   have different molecular masses

☐ C   have different solubilities in the two phases

☐ D   have different retention times.   **(1 mark)**

(c) Suggest an instrument that could be used to detect and identify the
components in step 4.

....   **(1 mark)**

Qualitative tests for functional groups (1)

1 Five molecules, **A**, **B**, **C**, **D** and **E**, are shown in the table.

A	B	C	D	E
(structure: butanoic acid)	(structure with OH groups)	(structure: butanal)	(structure: butan-2-one)	(structure with OH)

You are given five unlabelled samples, one of each substance. Use test-tube reactions to answer the questions.

(a) How could **A** be distinguished from the other four substances?

....

.... **(2 marks)**

(b) (i) Describe a single reaction that could distinguish **C** and **D** from the four remaining samples.

....

....

.... **(2 marks)**

(ii) How can **C** be distinguished from **D**?

....

....

.... **(2 marks)**

(c) Describe how **B** and **E** can be distinguished.

....

....

.... **(3 marks)**

(d) If molecules **A** and **B** are mixed with an acid catalyst, what is the number of carbons in the molecule of the product with the highest possible relative molecular mass?

☐ **A** 5

☐ **B** 10

☐ **C** 15

☐ **D** 20 **(1 mark)**

Qualitative tests for functional groups (2)

1 A compound **A** has the molecular formula $C_5H_9O_2Br$. When sodium carbonate solution is added effervescence of gas **B** is seen. If **A** is warmed with a mixture of silver nitrate solution and ethanol a precipitate **C** is formed. **A** can be warmed with sodium hydroxide solution to form **D**. When **D** is warmed with acidified potassium dichromate(VI) solution the mixture changes colour. **A** has a chiral centre. The carbon NMR spectrum of **A** has four peaks.

(a) (i) Identify gas **B**.

....

(1 mark)

   (ii) The formation of gas **B** shows possible functional group(s) in **A** of

> Consider whether all of these functional groups would fit the molecular formula.

     1   alcohol

     2   carboxylic acid

     3   acyl chloride

☐   A   only 1

☐   B   only 2

☐   C   only 3

☐   D   only 2 and 3   **(1 mark)**

(b) Give the formula and colour of precipitate **C**.

.... **(2 marks)**

(c) (i) What **type** of reaction has occurred converting **A** to **D**?

.... **(1 mark)**

   (ii) What is the colour change when **D** is warmed with acidified potassium dichromate(VI) solution?

.... **(1 mark)**

   (iii) Name the functional group that is present in **D** but not in **A**.

.... **(2 marks)**

(d) Draw the structure of **A** below.

> Start with the chiral centre and draw on the groups you know are present, then look at the molecular formula.

(2 marks)

Carbon-13 NMR spectroscopy

Use the data sheet to help answer the questions.

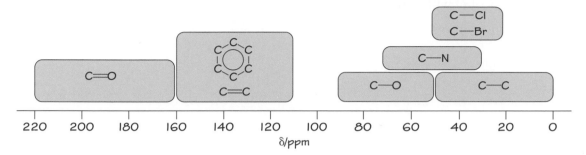

1 The carbon-13 NMR spectrum of an organic molecule is shown.

(a) What can be deduced from this spectrum?

1   The molecule contains five carbon atoms.

2   There are five different carbon environments in the molecule.

3   The molecule is unlikely to have a C=C bond.

☐ A   only 1 and 2

☐ B   only 2 and 3

☐ C   only 1 and 3

☐ D   1, 2 and 3   **(1 mark)**

(b) The molecule is an ester. The empirical formula of the molecule is C_3H_6O.

Draw the structure of a molecule that would give the number of peaks in the spectrum shown.

> Remember that the number of peaks indicates the number of different carbon environments. You only need to give one possible molecule – there are several possibilities.

(2 marks)

 Guided

(c) Draw a molecule of phenol and use it to explain how many peaks would be found in the carbon-13 NMR spectrum, with their likely chemical shift.

The number of different carbon environments is .... and these

are labelled on the diagram 1, 2 .... **(4 marks)**

Proton NMR spectroscopy

1   The proton NMR spectrum to the right is
    of an alcohol with three carbon atoms.

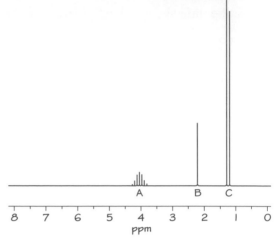

(a)   State the number of peaks that would be found in the proton NMR spectrum
      of each possible alcohol with their carbon atoms per molecule and hence
      identify the alcohol whose spectrum is shown.

      ....

      ....   **(3 marks)**

>Guided>   (b)   Explain which hydrogens cause each peak, **A**, **B** and **C** by considering the
              splitting patterns and give the ratio of the peak areas, **A : B : C**.

      Peak **A** is split into .... so has .... neighbours and so is ....

      Peak **B** is unsplit ....

      Peak **C** ....

      Ratio A : B : C is ....   **(7 marks)**

2   Cyclohexene, C_6H_{10}, reacts with hydrogen in the presence of a catalyst to form
    cyclohexane.

    (a)   Draw the structural formula of cyclohexene, showing all atoms.

                                                                                          (1 mark)

    (b)   State the number of peaks, and relative peak areas in the proton NMR
          spectrum for.

          (i)   cyclohexane:

          ....   **(1 mark)**

          (ii) cyclohexene:

          ....

          ....   **(2 marks)**

Identifying the structure of a compound from a proton (H-1) NMR spectrum

1 The proton NMR spectrum of the molecule bromoethane is shown.

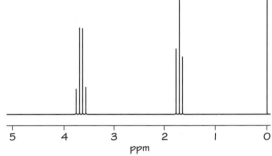

(a) The peak to the right is

☐   A   due to the CH_2 protons

☐   B   due to the CH_3 protons

☐   C   due to the bromine

☐   D   due to the TMS.

> Look at the shift value of the peak.

(1 mark)

(b) Explain the splitting pattern of the two left-hand peaks.

....

.... **(2 marks)**

(c) A spectrum of bromopropane is shown.

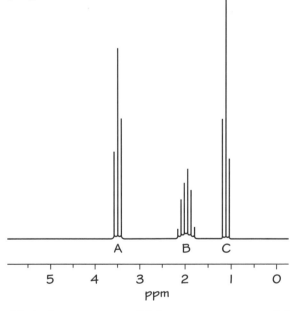

>**Guided**

(i) Explain whether this is the spectrum of 1-bromopropane or 2-bromopropane.

There are .... peaks ....

.... **(2 marks)**

(ii) Describe which protons cause the central peak.

.... **(1 mark)**

(iii) The **other** two peaks have the same splitting pattern. Describe how further information from the spectrum allows them to be allocated to the correct protons.

....

.... **(2 marks)**

Predicting a proton NMR spectrum

1   Consider the molecules below.

P	Q
CH₃ (attached to benzene ring)	H₃C–C(CH₃)₂–CH₂–CH₃ arrangement with CH₃ groups
R	**S**
H–C–C–C–C–H chain with H atoms above and below each carbon	H₃C–Si(CH₃)–CH₃ with CH₃ group above and CH₃ below

(a) Which of these is used as a standard for chemical shift measurements?

☐ **P**   ☐ **Q**   ☐ **R**   ☐ **S**   **(1 mark)**

(b) The number of peaks in the carbon-13 NMR spectrum is

	P	**Q**	**R**	**S**
☐ A	5	2	1	1
☐ B	5	4	2	1
☐ C	2	4	5	1
☐ D	2	2	5	1

                                                                          (1 mark)

(c) The number of different hydrogen environments in **P** is

☐ A 2   ☐ B 3   ☐ C 4   ☐ D 5   **(1 mark)**

(d) The molecule which would **not** have a singlet in the proton NMR spectrum is

☐ **P**   ☐ **Q**   ☐ **R**   ☐ **S**   **(1 mark)**

2   In the figure below, the upper proton NMR spectrum is that of ethanol, and the lower is that of ethanol mixed with D₂O.

Explain the similarities and differences in the two spectra.

....

....

....

....

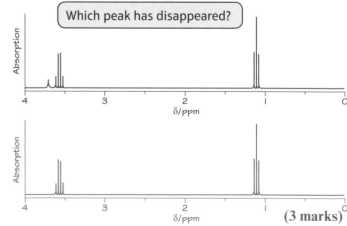

Which peak has disappeared?

                                                                          (3 marks)

Deducing the structure of a compound from a range of data

1   A compound contains 62.1% carbon, 27.6% oxygen and hydrogen only and has only one functional group. Its mass spectrum has a molecular ion at 116. The IR spectrum has a peak at $1710\,cm^{-1}$ and a broad peak at $2500-3300\,cm^{-1}$. The proton NMR spectrum has four peaks, **A**, **B**, **C** and **D**. The peak areas are in the ratio $2:3:1:6$ and are a quartet, triplet, singlet and singlet respectively.

(a)   (i)   Calculate the empirical formula of the molecule.

....

....

.... **(3 marks)**

> Guided

     (ii)   State and explain the molecular formula of the molecule.

The molecular ion ....

.... **(2 marks)**

(b)   Use the data on page 75 to explain the deduction that can be made from the IR spectrum.

....

.... **(2 marks)**

(c)   Use all of the information to draw and name the molecule.

> The broad peak in the IR spectrum and the number of oxygen atoms are a good starting point.

     Name: ....   **(2 marks)**

(d)   Explain which of the protons lead to each of the peaks in the NMR spectrum.

A: .... **(1 mark)**

B: .... **(1 mark)**

C: .... **(1 mark)**

D: .... **(1 mark)**

Exam skills 13

1 This question is about an aliphatic organic compound, **T**, which has one functional group and seven carbon atoms in each molecule.

The following observations are made when **T** is tested.

Test		Observation
1	**T** is warmed by acidified potassium dichromate(VI)	mixture turns from orange to green
2	**T** is warmed with Tollens' reagent	no visible change
3	**T** is heated under reflux with acidified potassium dichromate(VI). The reaction mixture is neutralised. Some solid sodium carbonate is added.	no effervescence

(a) What can be deduced from Test 1?

....

....

(3 marks)

> Be precise about the possible functional groups.

(b) What can be deduced from Test 2?

.... **(1 mark)**

(c) Explain what can be deduced from Test 3.

....

....

(3 marks)

> An **explanation** is needed, so explain how the evidence is used.

(d) What overall deduction can be made using Tests 1, 2 and 3?

.... **(1 mark)**

(e) The proton NMR spectrum of **T** has the following peaks:

Peak	Area	Splitting pattern
A	1	singlet
B	1	triplet
C	2	quintet
D	3	triplet
E	9	singlet

Draw the formula of molecule **T** and circle and label each group of hydrogens A to E.

(4 marks)

AS Level Chemistry
Practice paper 1

Breadth in chemistry

Time allowed: 1 hour 30 minutes

You must have:
• the data sheet for Chemistry A

You may use:
• a scientific calculator

INSTRUCTIONS
• Use black ink. You may use an HB pencil for graphs and diagrams.
• Answer all the questions.

INFORMATION
• The total marks for this paper is **70**.
• The marks for each question are shown in brackets [ ].

SECTION A

You should spend a maximum of 25 minutes on this section.

Answer **all** the questions.

1 A fluoride ion has a mass number of 19.

What is the number of electrons, protons and neutrons in the F^- ion?

	electrons	protons	neutrons
A	8	9	10
B	9	9	10
C	10	10	9
D	10	9	10

Your answer ☐ [1]

2 What is the formula of iron(III) oxide?

A FeO_3
B Fe_3O
C Fe_2O_3
D Fe_3O_2

Your answer ☐ [1]

3 Which of the following are properties of ionic compounds?

1: They have high melting points
2: Some dissolve in polar solvents
3: They conduct electricity only when molten

A 1, 2 and 3
B only 1 and 2
C only 2 and 3
D only 1

Your answer ☐ [1]

4 Which molecule has the correct bond angle?

A H_2O 180°
B NH_3 120°
C CH_4 109.5°
D BCl_3 107°

Your answer ☐ [1]

5 Which symbols for three elements are in order of increasing first ionisation energy?

A He Li Be
B Li Be B
C Be B C
D B C N

Your answer ☐ [1]

6 When some calcium oxide is added to a boiling tube of water, what happens?

1: a white precipitate forms
2: the mixture forms with a pH of approximately 11–12
3: hydrogen gas is evolved

A 1, 2 and 3
B only 1 and 2
C only 2 and 3
D only 1

Your answer ☐ [1]

7 Which of the following does not result in a white precipitate?

A sodium sulfate and barium nitrate solutions are mixed
B ammonium chloride and sodium hydroxide solutions are mixed
C carbon dioxide is bubbled through calcium hydroxide solution
D silver nitrate and potassium chloride solutions are mixed

Your answer ☐ [1]

8 How can the equilibrium yield of ammonia be increased?

$$N_2 + 3H_2 \rightleftharpoons 2NH_3 \qquad \Delta H = -92\,kJ\,mol^{-1}$$

A reducing the pressure
B reducing the temperature
C adding a catalyst
D increasing the temperature

Your answer ☐ [1]

9 Use the data to calculate the enthalpy change of reaction, in $kJ\,mol^{-1}$, for

$$H_2 + Br_2 \rightarrow 2HBr$$

Bond	Average bond enthalpy / $kJ\,mol^{-1}$
H–H	432
H–Br	364
Br–Br	193

A −51.5
B −103
C +103
D +261

Your answer ☐ [1]

10 In a Boltzmann distribution for a gas sample, if the temperature of the gas is increased

1: the area under the curve increases
2: the peak of the curve rises
3: the proportion of the curve for particles with energy ≥ activation energy increases

Which statement(s) are correct?

A only 1 and 2
B only 1 and 3
C only 2 and 3
D only 3

Your answer ☐ [1]

11 An organic hydrocarbon has the empirical formula CH_2. The molecular formula of the substance could be

1: CH_2
2: C_3H_6
3: $CH_3(CH_2)_{10}CH_3$

Which statement(s) are correct?

A only 1
B only 2
C only 3
D only 1 and 2

Your answer ☐ [1]

12 Which of the following has the greatest number of moles of the substance given?

The molar volume of a gas is $24\,dm^3$. The Avogadro constant is $6.0 \times 10^{23}\,mol^{-1}$.

A Cl_2 in $12\,dm^3$ Cl_2 gas
B H_2SO_4 in $250\,cm^3$ of $2.0\,mol\,dm^{-3}$ solution
C $CaCO_3$ in $75\,g$ of $CaCO_3$
D Ar in 3.0×10^{23} argon atoms

Your answer ☐ [1]

13 Examples of a weak acid are

1: $0.100\,mol\,dm^{-3}$ HCl
2: $2.75\,mol\,dm^{-3}$ H_2SO_4
3: $0.750\,mol\,dm^{-3}$ CH_3COOH

Which statement(s) are correct?

A only 1
B only 2
C only 3
D only 1 and 3

Your answer ☐ [1]

14 The IUPAC name of $CH_3CH(OH)CH_2CH_2OH$ is

A butane-1,3-diol
B 1,3-butanol
C 2,4-dihydroxybutane
D but-2,4-dianol

Your answer ☐ [1]

15 Which of the following molecules have at least one carbon atom with atoms arranged around the atom in a tetrahedral shape?

1: C_2H_6
2: C_2H_4
3: CH_3COOH

Which statement(s) are correct?

A only 1 and 2
B only 2 and 3
C only 1 and 3
D 1, 2 and 3

Your answer ☐ [1]

16 When an alkane is burnt in a limited supply of air, one product that is not formed is

A CO
B CO_2
C H_2
D H_2O

Your answer ☐ [1]

17 Consider the two molecules.

These molecules are not correctly described as

A E/Z isomers
B hydrocarbons
C stereoisomers
D saturated

Your answer ☐ [1]

18 Alkenes cannot undergo addition reactions with

   A hydrogen
   B potassium dichromate
   C bromine
   D steam

   Your answer ☐   **[1]**

19 A secondary alcohol can form

   A an alkene by elimination
   B an aldehyde by oxidation
   C a haloalkane by addition
   D a carboxylic acid by oxidation

   Your answer ☐   **[1]**

20 The haloalkane which reacts fastest with aqueous alkali is

   A CH_3CH_2F
   B CH_3CH_2Br
   C CH_3CH_2Cl
   D CH_3CH_2I

   Your answer ☐   **[1]**

SECTION B

Answer all the questions.

1 Chlorine is found is Group 7 of the periodic table.

  (a) Give the electronic configuration of an atom of chlorine.

  .... **[1]**

  (b) Draw a dot and cross diagram of a molecule of chlorine, showing outer electrons only.

        [2]

  (c) Chlorine exists as two isotopes. 75% of chlorine atoms are 35-chlorine and 25% are 37-chlorine.

    (i) What are isotopes?

    ....

    .... **[2]**

    (ii) Calculate the relative atomic mass of chlorine.

    .... **[2]**

    (iii) A sample of chlorine is placed in a mass spectrometer. Explain how the electron gun works, and give the formula of the particle formed when the electron gun acts on a sample of chlorine gas.

    .... **[3]**

  (d) Describe how you could test some sodium chloride to demonstrate the presence of chloride ions.

  ....

  ....

  .... **[4]**

  (e) In an experiment to test the reactivity of halogens, chlorine is bubbled into solutions of potassium halides. The results are:

	Br^- (aq)	I^- (aq)
Cl_2 (aq)	orange solution	brown solution

    (i) Identify the particle that causes the orange colour.

    .... **[1]**

    (ii) Write an ionic equation for the reaction of chlorine with iodide ions.

    .... **[2]**

    (iii) Suggest and explain what would you observe, if anything, if chlorine were bubbled into potassium fluoride solution.

    ....

    .... **[2]**

---

2 (a) Write down the equation, including state symbols, for the reaction whose enthalpy change is the standard enthalpy of formation of propane, C_3H_8.

  .... **[2]**

  (b) Use the enthalpy of combustion data to calculate the enthalpy change for the reaction given in (a).

	$\Delta_c H^\circ$/ kJ mol^{-1}
C(s)	−394
H_2(g)	−286
C_3H_8(g)	−2220

  $\Delta_f H^\circ$= .... kJ mol^{-1} **[3]**

  (c) Propane gas is used in portable heaters. The amount of energy released when one mole of propane gas is used in one of these heaters is measured. Suggest why the amount of heat released is less than 2220 kJ.

  .... **[1]**

3 A haloalkane has the molecular formula $C_3H_6Cl_2$.

  (a) Draw the structures of, and name, two isomers with this formula.

        [4]

  (b) When a haloalkane is warmed with aqueous potassium hydroxide, a nucleophilic substitution reaction occurs.

    (i) Write the ionic equation, using structural formulae, for the reaction of 2-bromobutane with hydroxide ions.

        [2]

    (ii) Identify the nucleophile in this reaction.

    .... **[1]**

    (iii) Give the functional group of the organic product.

    .... **[1]**

  (c) Explain why the release of haloalkanes into the environment is a health concern.

  ....

  ....

  .... **[3]**

---

4 Hydrated copper(II) sulfate, $CuSO_4.xH_2O$, is an ionic compound.

  (a) Describe an ionic lattice.

  ....

  .... **[2]**

  (b) Explain why copper(II) sulfate has to be melted or dissolved for it to be able to conduct electricity.

  ....

  .... **[1]**

  (c) 2.50 g of hydrated copper(II) sulfate is strongly heated. A white solid is formed and steam is given off. The white solid is weighed. Heating is repeated until there is no further loss in mass.

    (i) Identify the white solid.

    .... **[1]**

    (ii) Explain why the heating is repeated.

    .... **[1]**

    (iii) The final mass of the white solid was 1.60 g. Find the value of x in the formula of the hydrated copper(II) sulfate.

    ....

    ....

    .... **[4]**

5 Calcium carbonate, $CaCO_3$, reacts with hydrochloric acid to form calcium chloride, $CaCl_2$.

  (a) Write the equation for this reaction.

  .... **[2]**

  (b) Calculate the atom economy of this reaction to form calcium chloride.

  ....

  ....

  .... atom economy = ...% **[3]**

AS Level Chemistry
Practice paper 2

Depth in chemistry

Time allowed: 1 hour 30 minutes

You must have:
• the data sheet for Chemistry A

You may use:
• a scientific calculator

INSTRUCTIONS
• Use black ink. You may use an HB pencil for graphs and diagrams.
• Answer all the questions.

INFORMATION
• The total mark for this paper is **70**.
• The marks for each question are shown in brackets [ ].

Answer all the questions.

1 The carbonate, MCO_3, of an unknown group II metal, **M**, is heated in a crucible. The carbonate decomposes to form the oxide, MO, and carbon dioxide. The mass of the crucible and its contents is measured before and after heating. Further heating occurs, and the mass is measured again. The results are shown.

Mass of empty crucible/ g	9.552
Mass of crucible and MCO_3/ g	12.729
Mass of crucible and contents after first heating/ g	11.782
Mass of crucible and contents after second heating/ g	11.782

(a) Write an equation, using M for the group II metal, for the decomposition reaction.
....
[1]

(b) What is the reason for heating the crucible and contents for the second time?
....
[1]

(c) Calculate the moles of carbon dioxide formed in the reaction.
....
.... amount of carbon dioxide = ... mol
[2]

(d) (i) Calculate the mass of the metal oxide, MO, formed in the reaction.
.... mass = ... g
[1]

(ii) Use your answers to parts (c) and (d)(i) to calculate the relative atomic mass of M, and hence identify M.
....
.... A_r(M) = ... M is ....
[4]

(e) Suggest a reason why the loss in mass in such an experiment might be higher than the mass calculated from the chemical equation.
....
[1]

2 In an experiment to determine the enthalpy change of neutralisation when potassium hydroxide solution is mixed with dilute hydrochloric acid, the following apparatus is used.

25.0 cm³ of 2.0 mol dm⁻³ dilute hydrochloric acid is accurately measured and placed in the cup. 25.0 cm³ of 2.0 mol dm⁻³ sodium hydroxide solution is accurately measured and added to the cup. The mixture is stirred and the maximum temperature recorded. The results are shown.

Initial temperature of solutions/ °C	14.8
Maximum temperature of mixture/ °C	27.9

(a) Hydrochloric acid is a strong acid. What is meant by a strong acid?
....
[1]

(b) Write the ionic equation for the neutralisation reaction, including state symbols.
....
[2]

(c) (i) Use the formula $q = mc\Delta T$ to calculate the enthalpy change in the reaction in kJ. $c = 4.18\,J\,K^{-1}\,mol^{-1}$.
....
....
.... q = ... kJ
[2]

(ii) Calculate the moles of sodium hydroxide used.
....
.... amount of sodium hydroxide = ... mol
[2]

(d) (i) Define the term enthalpy change of neutralisation.
....
....
[1]

(ii) Use your answers to part (c)(i) and (ii) to find the enthalpy change of neutralisation, in kJ mol⁻¹, to three significant figures.
....
.... $\Delta_{neut}H$ = ... kJ mol⁻¹
[2]

(e) Suggest two improvements to the experimental method described.
....
....
[2]

3 The melting points of some elements are given, in °C.

magnesium	650
aluminium	660
silicon	1414
phosphorus	44
sulfur	115

(a) Explain, in terms of its structure and bonding, why silicon has a very high melting point.
....
....
....
[3]

(b) Explain, in terms of structure and bonding, why sulfur has a higher melting point than phosphorus. ....
....
[2]

(c) Magnesium and aluminium are very good conductors of electricity. Explain why, in terms of their structure.
....
....
[2]

(d) The graph below shows the volume of hydrogen evolved when an excess of magnesium is added to 100.0 cm³ of 1.00 mol dm⁻³ dilute hydrochloric acid at 20°C.

(i) Explain why the reaction would be faster at 35°C.
....
....
....
[3]

(ii) Sketch on the same axes the graph produced when an excess of magnesium is added to 100.0 cm³ of 2.00 mol dm⁻³ dilute hydrochloric acid at 35°C.
[1]

(iii) One way to calculate the amount of gas evolved in a reaction is by measuring the mass loss as the gas escapes. Suggest why this method is unlikely to be suitable for this experiment if typical laboratory apparatus is used.
....
....
[1]

4 (a) Draw the skeletal formula of a molecule of propene.

[1]

(b) Propene contains σ and π bonds.
(i) Explain what is meant by a σ bond.
....
[2]

(ii) Circle on your skeletal formula in (a) the π bond.
[1]

(c) Propene reacts with hydrogen bromide.
(i) Give the name of the two possible organic products.
....
[2]

(ii) Draw and name the mechanism for the reaction, to form one of the products that you have named in (i).

Mechanism ....
[5]

5 An experiment is set up to oxidise an alcohol. The alcohol is heated under reflux with potassium dichromate(VI) and one other reagent.
(a) Identify the other reagent needed for oxidation to occur.
....
[1]

(b) What colour change is seen if oxidation occurs?
....
[2]

(c) The structure of three alcohols is given. Identify each as primary, secondary or tertiary, and give the name of the product formed in the conditions listed above. If no reaction occurs, state 'no reaction'.
[6]

Structure	Primary, secondary or tertiary	Oxidation product
OH		
OH		
OH		

(d) Ethanol is a liquid that dissolves in water. Alkanes of similar relative molecular mass are gases that do not dissolve in water.
(i) Give the name of the type of intermolecular bonding found in ethanol that is not found in alkanes.
....
[1]

(ii) Explain why ethanol has this type of bonding but alkanes do not.

....

....   [2]

6   This question is about chlorine.
    (a) An old method to produce chlorine is the reaction of hydrogen chloride, HCl,
        with oxygen to make chlorine and water.
        (i) Construct an equation for this reaction.

        ....   [2]
        (ii) In such a reaction, 1.00 kg of hydrogen chloride formed 775 g chlorine.
             Calculate the percentage yield of the reaction.

             ....

             ....

             .... yield = ... %   [3]
        (iii) The reaction is carried out at about 400 °C. Calculate the volume of 775 g
              chlorine at this temperature and 10^5 Pa. $R = 8.31$ J K^{-1} mol^{-1}. Give your
              answer to three significant figures.

              ....

              ....

              .... volume = ... m^3 [4]
    (b) Give the electronic configuration of a chlorine atom.

        ....   [1]
    (c) Chlorine reacts with methane to form CCl$_4$.
        (i) State why a C–Cl bond is polar.

        ....   [1]
        (ii) Explain why the CCl$_4$ molecule is not polar.

        ....   [1]

5

A Level Chemistry
Practice paper 1

Periodic table, elements and physical chemistry

Time allowed: 2 hours 15 minutes

You must have:
• the data sheet for Chemistry A
You may use:
• a scientific calculator

INSTRUCTIONS
• Use black ink. You may use an HB pencil for graphs and diagrams.
• Answer all the questions.

INFORMATION
• The total mark for this paper is **100**.
• The marks for each question are shown in brackets [ ].
• Quality of extended responses will be assessed in questions marked with an asterisk (*).

SECTION A

You should spend a maximum of 20 minutes on this section.

Answer **all** the questions.

1 What is the formula of iron(III) oxide?
   A FeO_3
   B Fe_3O
   C Fe_2O_3
   D Fe_3O_2
   Your answer ☐   [1]

2 Which molecule has the correct bond angle?
   A H_2O 180°
   B NH_3 120°
   C CH_4 109.5°
   D BCl_3 107°
   Your answer ☐   [1]

3 Which symbols for three elements are in order of increasing first ionisation energy?
   A He, Li, Be
   B Li, Be, B
   C Be, B, C
   D B, C, N
   Your answer ☐   [1]

4 Which of the following does **not** result in a white precipitate?
   A sodium sulfate and barium nitrate solutions are mixed
   B ammonium chloride and sodium hydroxide solutions are mixed
   C carbon dioxide is bubbled through calcium hydroxide solution
   D silver nitrate and potassium chloride solutions are mixed
   Your answer ☐   [1]

5 How can the equilibrium yield of ammonia be increased?

$N_2 + 3H_2 \rightleftharpoons 2NH_3$ $\Delta H = -92\,kJ\,mol^{-1}$

   A reducing the pressure
   B reducing the temperature
   C adding a catalyst
   D increasing the temperature
   Your answer ☐   [1]

6 Use the data to calculate the enthalpy change of reaction, in $kJ\,mol^{-1}$, for

$H_2 + Br_2 \rightarrow 2HBr$

Bond	Average bond enthalpy / $kJ\,mol^{-1}$
H–H	432
H–Br	364
Br–Br	193

   A −51.5
   B −103
   C +103
   D +261
   Your answer ☐   [1]

7 In a Boltzmann distribution for a gas sample, if the temperature of the gas is increased
   1: the area under the curve increases
   2: the peak of the curve rises
   3: the proportion of the curve for particles with energy ≥ activation energy increases

   Which statement(s) are correct?

   A only 1 and 2
   B only 1 and 3
   C only 2 and 3
   D only 3
   Your answer ☐   [1]

8 What could the axes on the graph below be?

	y-axis	*x*-axis
A	pH of weak acid reactant	volume of added alkali
B	rate constant	temperature
C	rate	concentration of zero order reactant
D	rate	concentration of first order reactant

   Your answer ☐   [1]

9 What is the pH of a mixture of $25\,cm^3\ 0.010\,mol\,dm^{-3}$ NaOH and $50\,cm^3\ 0.010\,mol\,dm^{-3}$ HCl?
   A −2.00
   B 0.00
   C 2.47
   D 3.60
   Your answer ☐   [1]

10 A reaction may **not** occur because
   1: the activation energy is too high
   2: ΔG for the reaction is positive
   3: $E^{\bullet}$ for the reaction is positive

   Which statement(s) are correct?

   A only 1 and 2
   B only 2
   C only 2 and 3
   D only 3
   Your answer ☐   [1]

11 Which of the compounds would have the most exothermic lattice enthalpy?
   A NaBr
   B NaCl
   C MgF_2
   D $MgCl_2$
   Your answer ☐   [1]

12 *cis*-platin is the complex $Pt(NH_3)_2Cl_2$.

   Which statement about *cis*-platin is incorrect.

   A *cis*-platin and *trans*-platin are stereoisomers
   B *cis*-platin would bleach damp litmus paper
   C *cis*-platin is used as an anti-cancer drug
   D *cis*-platin and *trans*-platin have a square planar shape
   Your answer ☐   [1]

13 What is the enthalpy change for $CuSO_4(s) \rightarrow Cu^{2+}(aq) + SO_4^{2-}(aq)$?
   A enthalpy change of formation of copper(II) sulfate solution
   B enthalpy change of solution of copper(II) sulfate
   C enthalpy change of hydration of copper(II) sulfate
   D lattice enthalpy of copper(II) sulfate
   Your answer ☐   [1]

14 4.25 g of hydrated magnesium sulfate, $MgSO_4.xH_2O$, was strongly heated. The loss in mass was 2.18 g, leaving a residue found to be 0.0172 mol $MgSO_4$.
   What is the value of x?
   A 0.1
   B 2
   C 5
   D 7
   Your answer ☐   [1]

15 pK_{aw} of water is 12.70 at 75 °C. What is the pH of pure water at this temperature?
   A 1.10
   B 6.35
   C 7.00
   D 12.70
   Your answer ☐   [1]

SECTION B

Answer **all** the questions.

1 Chlorine is found in Group 7 of the periodic table.

(a) Give the electronic configuration of an atom of chlorine.

.... [1]

(b) In an experiment to test the reactivity of halogens, chlorine is bubbled into solutions of potassium halides. The results are:

	Br⁻ (aq)	I⁻ (aq)
Cl₂ (aq)	orange solution	brown solution

(i) Identify the particle that causes the orange colour.

.... [1]

(ii) Write an ionic equation for the reaction of chlorine with iodide ions.

.... [2]

(iii) Suggest, and explain, what would you observe, if anything, if chlorine were bubbled into potassium fluoride solution.

....

....

.... [2]

2 (a) Write down the equation, including state symbols, for the reaction whose enthalpy change is the standard enthalpy of formation of propane, C_3H_8.

.... [2]

(b) Use the enthalpy of combustion data to calculate the enthalpy change for the reaction given in part (a).

	$\Delta_c H^\ominus$/ kJ mol⁻¹
C(s)	−394
H₂(g)	−286
C₃H₈(g)	−2220

$\Delta_r H^\ominus$ = ... kJ mol⁻¹ [3]

(c) Propane gas is used in portable heaters. The amount of energy released when one mole of propane gas is used in one of these heaters is measured.

Suggest why the amount of heat released is less than 2220 kJ.

....

.... [1]

3 The carbonate, MCO_3, of an unknown Group II metal, M, is heated in a crucible. The carbonate decomposes to form the oxide, MO, and carbon dioxide. The mass of the crucible and its contents is measured before and after heating. Further heating occurs, and the mass is measured again. The results are shown.

Mass of empty crucible/ g	10.500
Mass of crucible and MCO₃/ g	12.729
Mass of crucible and contents after first heating/ g	11.782
Mass of crucible and contents after second heating/ g	11.782

(a) Write an equation, using M for the Group II metal, for the decomposition reaction.

.... [1]

(b) What is the reason for heating the crucible and contents for the second time?

.... [1]

(c) Calculate the moles of carbon dioxide formed in the reaction.

....

.... amount of carbon dioxide = ... mol [2]

(d) (i) Calculate the mass of the metal oxide, MO, formed in the reaction.

.... mass = ... g [1]

(ii) Use your answers to parts (c) and (d)(i) to calculate the relative atomic mass of M, and hence identify M.

.... Ar(M) = ... M is .... [3]

(e) Suggest a reason why the loss in mass in such an experiment might be higher than that the mass calculated from the chemical equation.

....

.... [1]

4 In an experiment to determine the enthalpy change of neutralisation when potassium hydroxide solution is mixed with dilute hydrochloric acid, the following apparatus is used.

25.0 cm³ of 2.0 mol dm⁻³ dilute hydrochloric acid is accurately measured and placed in the beaker. 25.0 cm³ of 2.0 mol dm⁻³ sodium hydroxide solution is accurately measured and added to the beaker. The mixture is stirred and the maximum temperature recorded. The results are shown.

Initial temperature of solutions/ °C	14.8
Maximum temperature of mixture/ °C	27.9

(a) Hydrochloric acid is a strong acid. What is meant by a strong acid?

....

.... [1]

(b) Write the ionic equation for the neutralisation reaction, including state symbols.

.... [2]

(c) (i) Use the formula $q = mc\Delta T$ to calculate the enthalpy change in the reaction in kJ.
$c = 4.18$ J K⁻¹ mol⁻¹.

....

....

.... q = ... kJ [2]

(ii) Calculate the moles of sodium hydroxide used.

....

.... amount of sodium hydroxide = ... mol [2]

(d) (i) Define the term enthalpy change of neutralisation.

....

.... [1]

(ii) Use your answers to parts (c)(i) and (ii) to find the enthalpy change of neutralisation, in kJ mol⁻¹, to three significant figures.

....

.... $\Delta_{neut}H$ = ... kJ mol⁻¹ [2]

(e) Suggest two improvements to the experimental method described.

....

.... [2]

5 (a) Magnesium is a very good conductor of electricity. Explain why, in terms of its structure.

....

.... [2]

(b) The graph below shows the volume of hydrogen evolved when an excess of magnesium is added to 100.0 cm³ of 1.00 mol dm⁻³ dilute hydrochloric acid at 20 °C.

(i) Explain why the reaction is faster at 35 °C.

....

....

.... [3]

(ii) Sketch on the same axes the graph produced when an excess of magnesium is added to 100.0 cm³ of 2.00 mol dm⁻³ dilute hydrochloric acid at 35 °C. [2]

(iii) One way to calculate the amount of gas evolved in a reaction is by measuring the mass lost as the gas escapes. Suggest why this method is unlikely to be suitable for this experiment if typical laboratory apparatus is used.

....

.... [1]

6 Some standard electrode potentials are given, in volts.

$Cl_2(g) + 2e^- \rightarrow 2Cl^-(aq)$	+ 1.36
$Br_2(l) + 2e^- \rightarrow 2Br^-(aq)$	+1.07
$Fe^{3+}(aq) + e^- \rightarrow Fe^{2+}(aq)$	+0.77
$I_2(aq) + 2e^- \rightarrow 2I^-(aq)$	+0.54
$VO^{2+}(aq) + 2H^+(aq) + e^- \rightarrow V^{3+}(aq) + H_2O(l)$	+0.34
$V^{3+}(aq) + e^- \rightarrow V^{2+}(aq)$	−0.26
$Fe^{2+}(aq) + 2e^- \rightarrow Fe(s)$	−0.44

(a) State why the reaction of Fe^{3+} ions to form iron is described as reduction.

....

.... [1]

(b) These electrode potentials are measured using a standard hydrogen electrode. Describe a standard hydrogen electrode and give its $E^\ominus$ value.

....

....

....

.... $E^\ominus$ = ... V [4]

(c) State and explain, in terms of the halogen atom, which halogen from the table is the most powerful oxidising agent, giving evidence from the table to support your answer.

....

....

....

.... [4]

(d) Give the oxidation number of vanadium in each species in the table.
VO^{2+} .... V^{3+} .... V^{2+} .... [2]

(e) Some iron is added to a solution containing VO^{2+} ions
(i) Write ionic equations for every reaction that will occur.

....

....

....

.... [4]

(ii) Calculate $E^\ominus$ values for each of the reactions in part (i).

.... [2]

7 When a solution containing Cu^{2+} ions has an excess of ammonia solution added, an intermediate stage in the overall reaction can be represented as follows.

$[Cu(H_2O)_6]^{2+} + 2NH_3 \rightleftharpoons [Cu(NH_3)_2(H_2O)_4]^{2+} + 2H_2O$

(a) (i) Describe what you would see as ammonia is added dropwise, with shaking, and then to excess.

....

.... [3]

(ii) What type of reaction is represented in the equation?

.... [1]

(b) An equilibrium constant can be written for this reaction, called the stability constant K_{stab}. The two free water molecules on the right-hand side are omitted.

(i) Write the expression for K_{stab}.

.... [1]

(ii) Give the units of K_{stab}.

.... [1]

(c) Aqueous copper(II) ions also react with the ligand 1,2 diaminoethane (en).
$[Cu(H_2O)_6]^{2+} + en \rightleftharpoons [Cu(en)(H_2O)_4]^{2+} + 2H_2O$
K_{stab} for the equilibrium in part (c) is larger than that for the equilibrium in part (b).
(i) The enthalpy change for each reaction is similar. Explain why.

....
....
[2]

(ii) Suggest, using the equation $\Delta G = \Delta H - T\Delta S$, why K_{stab} for the equilibrium in part (c) is larger.

....
....
[2]

8 Nitrogen monoxide reacts with oxygen to form nitrogen dioxide. The enthalpy change for this reversible reaction is $-115\,kJ\,mol^{-1}$.

$2NO(g) + O_2(g) \rightleftharpoons 2NO_2(g)$

(a) Explain what conditions would lead to the highest yield of nitrogen dioxide.

....
....
....
....
[4]

(b) In an experiment, 1.2 mol NO and 0.8 mol O_2 are mixed in a 2.0 dm³ container and left to reach equilibrium. At equilibrium, 75% of the NO has reacted.
(i) Write an expression for K_c.

.... [1]

(ii) Calculate the value of K_c, giving the units.

....
....
....
.... K_c = ... [4]

(c) An experiment was carried out to determine the order of each of the reactants.

Experiment	[NO]/ mol dm⁻³	[O₂]/ mol dm⁻³	Rate/ mol dm⁻³ s⁻¹
1	1.5×10^{-5}	0.5×10^{-5}	2.1×10^{-7}
2	4.5×10^{-5}	0.5×10^{-5}	1.9×10^{-6}
3	1.5×10^{-5}	2.0×10^{-5}	8.4×10^{-7}

(i) Determine the order with respect to each reactant.

....
....
[2]

(ii) State the rate equation for the reaction.

.... [1]

(iii) Calculate the rate constant, and give units, using the data for Experiment 1.

....
.... k = ... units .... [3]

(iv) The mechanism for the reaction is
Step 1: $2NO \rightarrow N_2O_2$
Step 2: $N_2O_2 + O_2 \rightarrow 2NO_2$
Explain which is the rate-determining step.

....
[2]

A Level Chemistry
Practice paper 2
Synthetic and analytical techniques

Time allowed: 2 hours 15 minutes

You must have:
• the data sheet for Chemistry A
You may use:
• a scientific calculator

INSTRUCTIONS
• Use black ink. You may use an HB pencil for graphs and diagrams.
• Answer all the questions.

INFORMATION
• The total mark for this paper is **100**.
• The marks for each question are shown in brackets [ ].
• Quality of extended responses will be assessed in questions marked with an asterisk (*).

SECTION A

You should spend a maximum of 20 minutes on this section.

Answer **all** the questions.

1 An organic hydrocarbon has the empirical formula CH_2. The molecular formula of the substance could be
  1: CH_2
  2: C_3H_6
  3: $CH_3(CH_2)_{10}CH_3$

Which statement(s) are correct?

  A   only 1
  B   only 2
  C   only 3
  D   only 1 and 2

Your answer ☐   [1]

2 Which of the following has the greatest number of moles of the substance given?

The molar volume of a gas is $24\,dm^3$. The Avogadro constant is $6.0 \times 10^{23}\,mol^{-1}$.

  A $12\,dm^3\ CH_4$ gas
  B CH_3COOH in $250\,cm^3$ of $2.0\,mol\,dm^{-3}$ solution
  C $75\,g$ of C_7H_{14}
  D 3.0×10^{23} carbon atoms

Your answer ☐   [1]

3 Which of the following molecules have at least one carbon atom with atoms arranged around the atom in a tetrahedral shape?
  1: C_2H_6
  2: C_2H_4
  3: CH_3COOH

Which statement(s) are correct?

  A   only 1 and 2
  B   only 2 and 3
  C   only 1 and 3
  D   1, 2 and 3

Your answer ☐   [1]

4 When an alkane is burnt in a limited supply of air, one product that is not formed is
  A   CO
  B CO_2
  C H_2
  D H_2O

Your answer ☐   [1]

5 Alkenes cannot undergo addition reactions with
  A   hydrogen
  B   potassium dichromate
  C   bromine
  D   steam

Your answer ☐   [1]

6 A secondary alcohol can form
  A   an alkene by elimination
  B   an aldehyde by oxidation
  C   a haloalkane by addition
  D   a carboxylic acid by oxidation

Your answer ☐   [1]

7 The haloalkane which reacts fastest with aqueous alkali is
  A CH_3CH_2F
  B CH_3CH_2Br
  C CH_3CH_2Cl
  D CH_3CH_2I

Your answer ☐   [1]

8 The name of the molecule with the skeletal formula shown is
  A   cyclohexanol
  B   hydroxybenzene
  C   phenol
  D   1-hydroxy cyclohexane

Your answer ☐   [1]

9 Which of the following is **not** a possible benefit of processing waste polymers?
  A   HCl gas formed can be used for acid production
  B   Combustion can release useful energy
  C   Breakdown products can be used to make other plastics
  D   Less space is used up in landfill sites

Your answer ☐   [1]

10 If benzene had a Kekulé structure
  1: All carbon–carbon bond lengths would be the same
  2: Benzene would readily undergo addition reactions
  3: The enthalpy change of hydrogenation would be less exothermic

Which statement(s) are correct?

  A   only 1
  B   only 2
  C   only 3
  D   1, 2 and 3

Your answer ☐   [1]

11 Which of the following will completely react with 1 mol of HCl in solution?
  A $1\,mol$ of CH_3CH_2OH
  B $0.5\,mol$ of $CH_3CH(NH_2)COOH$
  C $1\,mol$ of CH_3CONH_2
  D $0.5\,mol$ of $C_6H_4(NH_2)_2$

Your answer ☐   [1]

12 Evidence that a solid organic compound contains a small amount of impurity can be that
  1: The measured melting point is too high
  2: The IR spectrum does not contain the expected peaks for the compound
  3: The mass spectrum has peaks at unexpected values

Which statement(s) are correct?

  A   only 1 and 2
  B   only 1 and 3
  C   only 2 and 3
  D   only 3

Your answer ☐   [1]

13 A molecule, **G**, containing five carbon atoms, is shaken with bromine water and the mixture turns colourless. When reacted with steam in the presence of an acid catalyst, two isomers are formed. When warmed with acidified potassium dichromate, the mixture with one of these isomers turns from orange to green, and there is no visible change with the other isomer.

G could be

  A   pentan-1-ol
  B   pent-2-ene
  C   2-methyl but-2-ene
  D   3-methyl but-1-ene

Your answer ☐   [1]

14 The high resolution proton NMR spectrum of I–CH_2–CH_2–Br has
  A   1 peak that is not split
  B   1 peak that is a doublet
  C   2 peaks that are doublets
  D   2 peaks that are each triplets

Your answer ☐   [1]

15 Which molecule could give the carbon-13 NMR spectrum shown?

A

B

C

D

Your answer ☐   [1]

1

2

3

4

158

SECTION B

Answer all the questions.

1 A haloalkane has the molecular formula $C_3H_6Cl_2$.
   (a) Draw the structures of, and name, two isomers with this formula.

....   **[4]**

   (b) When a haloalkane is warmed with aqueous potassium hydroxide, a nucleophilic substitution reaction occurs.
      (i) Write the ionic equation, using structural formulae, for the reaction of 2-bromobutane with hydroxide ions.

                                                                                                [2]

      (ii) Identify the nucleophile in this reaction.

....   **[1]**

      (iii) Give the functional group of the organic product.

....   **[1]**

   (c) Explain why the release of haloalkanes into the environment is a health concern.

....
....
....   **[3]**

2 Propan-1-ol reacts with methanoic acid to form an ester.
   (a) (i) Write the equation for this reaction.

....   **[2]**

      (ii) What is used as a catalyst for this reaction?

....   **[1]**

      (iii) Draw the skeletal formula and give the name of the ester formed.

      Name: ....   **[2]**

   (b) Calculate the atom economy of this reaction to form the ester.

....
....
.... atom economy = ...%   **[3]**

   (c) Another ester, ethyl ethanoate, has the structure given.

      (i) Give the name of the shape of the atoms arranged around the carbonyl carbon and suggest a likely value for the C–C=O bond angle.

....   **[2]**

      (ii) What is the empirical formula of this molecule?

....   **[1]**

   (d) Two ways of making the ester ethyl ethanoate are from a carboxylic acid and an alcohol or from an acid anhydride and the same alcohol.
      (i) Name the reactants

         carboxylic acid: ....   **[1]**

         alcohol: ....   **[1]**

         acid anhydride: ....   **[1]**

      (ii) State and explain an advantage of using the acid anhydride route.

....
....   **[2]**

      (iii) State and explain a precaution that has to be taken when using the acid anhydride that does not have to be taken when using the carboxylic acid.

....
....   **[2]**

3 Aspirin, a widely used painkiller, can be made by reacting 2-hydroxybenzoic acid with ethanoic anhydride. A method is given.
   1. Weigh accurately in a quickfit flask about 5.0 g 2-hydroxybenzoic acid.
   2. Add 10.0 cm³ ethanoic anhydride to the flask.
   3. Add with care 15 drops of concentrated phosphoric(V) acid, swirling carefully.
   4. Heat under reflux for 5 minutes.
   5. After 5 minutes add 10 cm³ water down the condenser, being careful of the vigorous reaction.
   6. Pour the contents into a small beaker of cold water, and place the beaker in an ice bath.
   7. Filter off the solid aspirin using filtration under reduced pressure.
   8. Recrystallise the aspirin using a minimum volume of hot water.

Results

Mass of quickfit flask/ g	58.35
Mass of flask with 2-hydroxybenzoic acid/ g	63.28
Mass of recrystallised aspirin formed/ g	4.37

   (a) In step 2, the ethanoic acid can be measured using a measuring cylinder. Why is it not necessary to use a more precise 10.0 cm³ pipette?

....   **[1]**

   (b) State and explain a suitable safety precaution, in addition to wearing a laboratory coat and safety goggles, in step 3.

....   **[1]**

   (c) Draw a diagram of the apparatus used in step 4 to heat under reflux.

                                                                                                [3]

   (d) What causes the vigorous reaction in step 5, and why is this step carried out?

....   **[2]**

   (e) Draw a diagram of the apparatus used for filtration under reduced pressure in step 7.

                                                                                                [2]

   (f) (i) Suggest why aspirin is soluble in warm water.

....
....   **[1]**

      (ii) Explain why hot water is used in the recrystallisation in step 8.

....
....   **[2]**

      (iii) Explain why a minimum volume of hot water is used in the recrystallisation in step 8.

....
....   **[2]**

   (g) The skeletal formulae of 2-hydroxybenzoic acid and aspirin are given.

      (i) Name three functional groups found in either, or both, of these molecules.

....   **[3]**

      (ii) Using the data above, calculate the percentage yield of the experiment.

....
....
.... % yield = ...%   **[4]**

      (iii) Give one reason why the yield is less than 100%.

....   **[1]**

   (h) State and explain the value of a peak that is found in the IR spectrum of 2-hydroxybenzoic acid but not aspirin.

....   **[2]**

4 An organic compound contains 55.2 % carbon, 18.4% oxygen, 16.1% nitrogen and 10.3% hydrogen.
   (a) (i) Calculate the empirical formula of the compound.

....
....   **[3]**

      (ii) The relative molecular mass of the compound is 87. Give the molecular formula.

....   **[1]**

   (b) Give a structural isomer of this molecule that
      (i) rotates the plane of plane-polarised light

                                                                                                [1]

      (ii) contains a primary amine and an aldehyde group

                                                                                                [1]

      (iii) is an amide

                                                                                                [1]

(c) Describe a test-tube reaction that would allow you to distinguish between
     your isomers in parts (b)(ii) and (iii).

....

....

....   [3]

5 Consider the reaction sequence

     A   B   C   D

(a) Give the names of the four molecules:

     A: ....   [1]

     B: ....   [1]

     C: ....   [1]

     D: ....   [1]

(b) (i) What reagents are added in the conversion of **A** to **B**?

     ....   [2]

     (ii) What type of reaction is the conversion of **A** to **B**?

     ....   [1]

     (iii) Write the mechanism for the conversion of **A** to **B**, including the reaction
           of the reagents to form the species which then reacts with **A**.

                                                                                              [4]

(c) (i) What reagents are added in the conversion of **B** to **C**?

     ....   [2]

     (ii) What type of reaction is the conversion of **B** to **C**?

     ....   [1]

(d) In a mass spectrum,
     (i) Give the value of the peak for the molecular ion of molecule **A**.

     ....   [1]

     (ii) Give the value of a peak you might find in all of the spectra, **A**, **B**, **C** and **D**.

     ....   [1]

     (iii) Give the formula of a species that would give a fragmentation peak that
           you would find only in the spectrum of **B**.

     ....   [1]

9

A Level Chemistry
Practice paper 3

Unified chemistry

Time allowed: 1 hour 30 minutes

You must have:
• the data sheet for Chemistry A
You may use:
• a scientific calculator

INSTRUCTIONS
• Use black ink. You may use an HB pencil for graphs and diagrams.
• Answer all the questions.

INFORMATION
• The total mark for this paper is **70**.
• The marks for each question are shown in brackets [ ].
• Quality of extended responses will be assessed in questions marked with an asterisk (*).

Answer **all** the questions.

1 The following information is adapted from "Mineral Planning Factsheet: Limestone" produced by the British Geological Society.

Limestones are sedimentary rocks consisting principally of calcium carbonate, $CaCO_3$. Industrial limestones contain small amounts of impurities such as iron, sulfur, silica and lead. Industrial limestone is an important raw material, and it is also used to make lime, CaO, and then 'hydrated lime', $Ca(OH)_2$.

(a) (i) Write the equation for the formation of 'hydrated lime' from lime, including state symbols.

.... [2]

(ii) What would you observe if the reaction in (a)(i) was carried out in the laboratory?

.... [2]

(iii) Give a commercial use for 'hydrated lime'.

.... [1]

(b) When carbon dioxide is bubbled through 'hydrated lime' a white precipitate forms.
Write the equation for this reaction, including state symbols, and underline the substance that forms the white precipitate.

....
.... [3]

(c) (i) Suggest why, for some uses, it is important to know whether a sample of limestone contains impurities.

....
.... [2]

(ii) In an experiment, you are provided with 3.76 g limestone and 1.50 mol dm⁻³ hydrochloric acid. It is found that 48.90 cm³ of acid are required to completely react with the calcium carbonate. Calculate the percentage of calcium carbonate in the limestone.

Purity = ... % [4]

(iii) What assumption is made about the impurities in the above calculation?

.... [1]

(d) Calcium carbonate decomposes when heated.
$CaCO_3(s) \rightarrow CaO(s) + CO_2(g)$
(i) Write an expression for K_p.

.... [1]

(ii) Data for this reaction and the substances is given.

$\Delta H^\ominus$	+179 kJ mol⁻¹
$S^\ominus$ (CaCO$_3$)	92.9 J K⁻¹ mol⁻¹
$S^\ominus$ (CaO)	40.0 J K⁻¹ mol⁻¹
$S^\ominus$ (CO$_2$)	214.0 J K⁻¹ mol⁻¹

Calculate $\Delta S^\ominus$ for the reaction, and hence determine the minimum temperature at which the reaction is feasible.

$\Delta S^\ominus$ .... J K⁻¹ mol⁻¹   T = ... K [4]

2 The pH curve of a neutralisation reaction is shown.

(a) Describe how, in the laboratory, the data for this graph could be collected.

....
....
.... [3]

(b) Was the acid and was the alkali in the neutralisation reaction strong or weak?

Acid: .... Alkali:.............................. [1]

(c) A buffer solution is made by adding 20.0 cm³ 0.100 mol dm⁻³ potassium hydroxide solution to 100 cm³ 0.120 mol dm⁻³ dilute ethanoic acid, $K_a = 1.76 \times 10^{-5}$ mol dm⁻³. Calculate the pH of the buffer.

pH = ... [5]

3 Manganese dioxide, MnO_2, is a catalyst in the decomposition of hydrogen peroxide, H_2O_2.
(a) State the oxidation number of
(i) Mn in MnO_2 ....
(ii) O in H_2O_2 .... [2]
(b) Describe how a catalyst increases the rate of reaction.

....
.... [2]

(c) Manganese forms compounds which when dissolved in water contain the $[Mn(H_2O)_6]^{2+}$.
(i) Give the electronic configuration of a Mn^{2+} ion.

.... [1]

(ii) What colour is $[Mn(H_2O)_6]^{2+}$?

.... [1]

4 (a) Draw the skeletal formula of a molecule of propene.

[1]

(b) Propene contains σ and π bonds.
(i) Explain what is meant by a σ bond.

....
.... [2]

(ii) Circle on your skeletal formula in part (a) the carbon atom(s) around which there is a tetrahedral arrangement of bonds. [1]
(c) Propene reacts with hydrogen bromide.
(i) Give the name of the two possible organic products.

.... [2]

(ii) Draw and name the mechanism for the reaction, to form the major product that you have named in part (i).

Name of mechanism: .... [5]

---

5 This question is about chlorine.
(a) An old method to produce chlorine is the reaction of hydrogen chloride, HCl, with oxygen to make chlorine and water.
(i) Construct an equation for this reaction.

.... [2]

(ii) In such a reaction, 1.00 kg of hydrogen chloride formed 775 g chlorine. Calculate the percentage yield of the reaction.

....
....
.... yield = ... % [3]

(b) Chlorine forms a molecule CCl₄ with polar C–Cl bonds. Explain why the CCl₄ molecule is not polar.

.... [1]

6 You are given samples of a pure liquid **A** which has three carbon atoms and one functional group.

When sodium carbonate powder is added to **A**, effervescence of gas **B** is seen.

When sulfurous dichloride oxide is added to pure **A** pungent fumes of **C** are released and organic product **D** is formed.

When **D** is reacted with methanol, the organic compound **E** is formed and steamy fumes **F** are released. **E** has the molecular formula $C_4H_8O_2$.

(a) Give the formulae of compounds **A** to **F**, using structural formulae for all organic substances.

A: .... [1]
B: .... [1]
C: .... [1]
D: .... [1]
E: .... [1]
F: .... [1]

(b) (i) Suggest a structure of an isomer of **E** which is an aliphatic diol.

[1]

(ii) Identify the other functional group in the isomer given in (b)(i), apart from alcohol.

.... [1]

(c) The structure of a molecule of glutamic acid is shown.

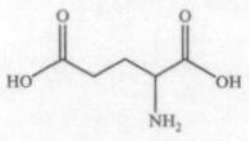

(i) Give the structural formula of the organic product when glutamic acid reacts with dilute hydrochloric acid.

[1]

(ii) Give the structural formula of the organic product when glutamic acid reacts with excess sodium hydroxide solution.

[1]

(iii) Glutamic acid is chiral. Circle the chiral centre on the diagram above. [1]

(d) Glutamic acid can be used to make the food additive monosodium glutamate (MSG), which also occurs naturally in foods such as tomatoes and mushrooms. Industrially, glutamic acid is made by fermentation of a plant such as cassava. The glutamic acid accumulates in the fermentation broth and is separated as an aqueous solution from the broth by filtration. The filtrate is concentrated by evaporation and purified by recrystallisation. It is then converted to MSG.

(i) Suggest one reason why MSG is manufactured by fermentation rather than by being extracted from tomatoes.

....

[1]

(ii) What is meant by **concentrated by evaporation**?

....

[1]

(iii) Describe the method of recrystallistion.

....

....

....

[3]

(iv) How would you show that the recrystallised glutamic acid was pure?

....

....

[2]

5

Periodic table

Group

	1 (1)	2 (2)	3 (3)	4 (4)	5 (5)	6 (6)	7 (7)	8 (8)	9 (9)	10 (10)	11 (11)	12 (12)	3 (13)	4 (14)	5 (15)	6 (16)	7 (17)	8 (18)
Period 1	1 **H** Hydrogen 1.0																	2 **He** Helium 4.0
Period 2	3 **Li** Lithium 6.9	4 **Be** Beryllium 9.0											4 **B** Boron 10.8	6 **C** Carbon 12.0	7 **N** Nitrogen 14.0	8 **O** Oxygen 16.0	9 **F** Fluorine 19.0	10 **Ne** Neon 20.2
Period 3	11 **Na** Sodium 23.0	12 **Mg** Magnesium 24.3											13 **Al** Aluminium 27.0	14 **Si** Silicon 28.1	15 **P** Phosphorus 31.0	16 **S** Sulfur 32.1	17 **Cl** Chlorine 35.5	18 **Ar** Argon 39.9
Period 4	19 **K** Potassium 39.1	20 **Ca** Calcium 40.1	21 **Sc** Scandium 45.0	22 **Ti** Titanium 47.9	23 **V** Vanadium 50.9	24 **Cr** Chromium 52.0	25 **Mn** Manganese 54.9	26 **Fe** Iron 55.8	27 **Co** Cobalt 58.9	28 **Ni** Nickel 58.7	29 **Cu** Copper 63.5	30 **Zn** Zinc 65.4	31 **Ga** Gallium 69.7	32 **Ge** Germanium 72.6	33 **As** Arsenic 74.9	34 **Se** Selenium 79.0	35 **Br** Bromine 79.9	36 **Kr** Krypton 83.8
Period 5	37 **Rb** Rubidium 85.5	38 **Sr** Strontium 87.6	39 **Y** Yttrium 88.9	40 **Zr** Zirconium 91.2	41 **Nb** Niobium 92.9	42 **Mo** Molybdenum 95.9	43 **Tc** Technetium (98)	44 **Ru** Ruthenium 101.1	45 **Rh** Rhodium 102.9	46 **Pd** Palladium 106.4	47 **Ag** Silver 107.9	48 **Cd** Cadmium 112.4	49 **In** Indium 114.8	50 **Sn** Tin 118.7	51 **Sb** Antimony 121.8	52 **Te** Tellurium 127.6	53 **I** Iodine 126.9	54 **Xe** Xenon 131.3
Period 6	55 **Cs** Caesium 132.9	56 **Ba** Barium 137.3	57 **La*** Lanthanum 138.9	72 **Hf** Hafnium 178.5	73 **Ta** Tantalum 180.9	74 **W** Tungsten 183.8	75 **Re** Rhenium 186.2	76 **Os** Osmium 190.2	77 **Ir** Iridium 192.2	78 **Pt** Platinum 195.1	79 **Au** Gold 197.0	80 **Hg** Mercury 200.6	81 **Tl** Thallium 204.4	82 **Pb** Lead 207.2	83 **Bi** Bismuth 209.0	84 **Po** Polonium (209)	85 **At** Astatine (210)	86 **Rn** Radon (222)
Period 7	87 **Fr** Francium (223)	88 **Ra** Radium (226)	89 **Ac*** Actinium (227)	104 **Rf** Rutherfordium (261)	105 **Db** Dubnium (262)	106 **Sg** Seaborgium (266)	107 **Bh** Bohrium (264)	108 **Hs** Hassium (277)	109 **Mt** Meitnerium (268)	110 **Ds** Darmstadtium (271)	111 **Rg** Roentgenium (272)	112 **Cn** Copernicium 112	114 **Fl** flerovium 114			116 **Lv** livermorium 116		

Lanthanides and actinides:

58 **Ce** Cerium 140.1	59 **Pr** Praseodymium 140.9	60 **Nd** Neodymium 144.2	61 **Pm** Promethium 144.9	62 **Sm** Samarium 150.4	63 **Eu** Europium 152.0	64 **Gd** Gadolinium 157.2	65 **Tb** Terbium 158.9	66 **Dy** Dysprosium 162.5	67 **Ho** Holmium 164.9	68 **Er** Erbium 167.3	69 **Tm** Thulium 168.9	70 **Yb** Ytterbium 173.0	71 **Lu** Lutetium 175.0
90 **Th** Thorium 232.0	91 **Pa** Protactinium (231)	92 **U** Uranium 238.1	93 **Np** Neptunium (237)	94 **Pu** Plutonium (242)	95 **Am** Americium (243)	96 **Cm** Curium (247)	97 **Bk** Berkelium (245)	98 **Cf** Californium (251)	99 **Es** Einsteinium (254)	100 **Fm** Fermium (253)	101 **Md** Mendelevium (256)	102 **No** Nobelium (254)	103 **Lr** Lawrencium (257)

Key

Atomic number
Atomic symbol
Name
Relative atomic mass

Data booklet

Physical constants

Avogadro constant (Na)	$6.02 \times 10^{23} \, \text{mol}^{-1}$
Gas constant (R)	$8.134 \, \text{J mol}^{-1} \, \text{K}^{-1}$
Molar volume of ideal gas:	
at r.t.p.	$24.0 \, \text{dm}^3 \, \text{mol}^{-1}$
Specific heat capacity of water (c)	$4.18 \, \text{J g}^{-1} \, \text{K}^{-1}$
Ionic product of water (K_w)	$1.00 \times 10^{-14} \, \text{mol}^2 \, \text{dm}^{-6}$ at 298 K

1 tonne = 10^6 g

Arrhenius equation $k = Ae^{-Ea/RT}$ or $\ln k = -Ea/RT + \ln A$

Infrared spectroscopy

Characteristic infrared absorptions in organic molecules

Bond	Wavenumber range/cm^{-1}
C—C Alkanes, alkyl chains	750–1100
C—X Haloalkanes (X—Cl, Br, I)	500–800
C—F Fluoroalkanes	1000–1300
N—H Amine, amide	3300–3500
O—H Alcohols and phenols Carboxylic acids	3200–3600 2500–3300 (broad)
C=C Alkenes Arene	1620–1680 Several peaks in range of 1450–1650 (variable)
C—O Alcohols, esters, carboxylic acids	1000–1350
C=O Aldehydes, ketones, carboxylic acids, acyl chlorides, esters, amides, acid anhydrides	1630–1820
Triple bond stretching vibrations CN CC	 2220–2260 2260–2100

^{1}H nuclear magnetic resonance chemical shifts relative to tetramethylsilane (TMS)

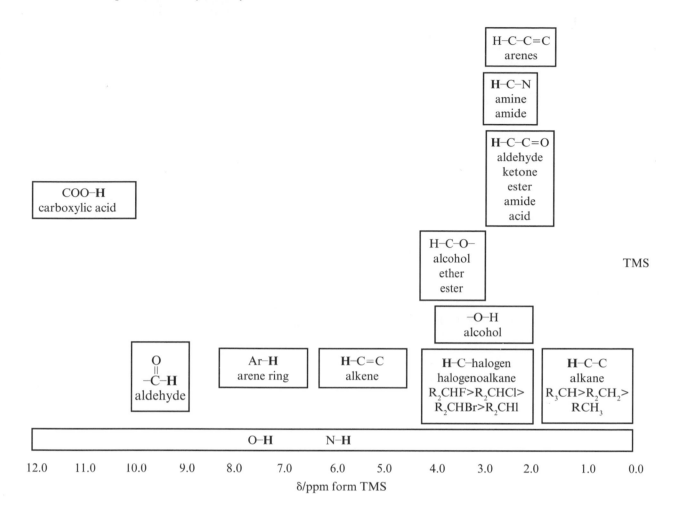

^{13}C nuclear magnetic resonance chemical shifts relative to tetramethylsilane (TMS)

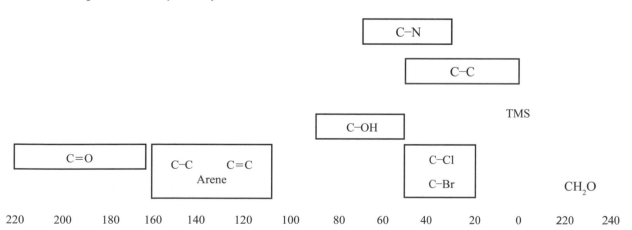

Chemical shifts are variable and can vary depending on the solvent, concentration and substituents. As a result, shifts may be outside the ranges indicated above.

OH and NH chemical shifts are very variable and are often broad. Signals are not usually seen as split peaks.

Note that CH bonded to 'shifting groups' on either side, e.g. O–CH$_2$–C=O, may be shifted more than indicated above.

Answers

You will find some advice next to some of the answers. This is written in *italics*. It is not part of the mark scheme but just gives you a little more information.

1. Atomic structure and isotopes

1. (a) atoms with the same number of protons / same atomic number **(1)** but different numbers of neutrons / different mass number **(1)**
   (b) one proton **(1)**, one electron **(1)**
   (c) the only atoms with no neutrons
   (d) **B**
   (e) (i) (reactions depend on the) number of electrons **(1)**, which is the same for each isotope **(1)**
      (ii) The molecules $^3H_2$ are heavier so would move more slowly **(1)** so reactions would be (slightly) slower. **(1)**

2. Relative masses

1. (a) Mass of an isotope **(1)** compared with $\frac{1}{12}$ th of the mass of a carbon-12 atom. **(1)**
   (b) (i) ^{35}Cl: 17 protons, 17 electrons, 18 neutrons; ^{37}Cl: 17 protons, 17 electrons, 20 neutrons *(4 marks, deduct one for each error)*
      (ii) This is the weighted mean mass of all of the isotopes of the element compared with $\frac{1}{12}$ th of the mass of a carbon-12 atom. **(1)**
      (iii) More atoms of chlorine are ^{35}Cl than ^{37}Cl.
2. 3
3. (a) 71
   (b) 187.5
   (c) 237.9

3. Using mass spectroscopy

1. (a) (i) $^{79}Br^+$
      (ii) Bromine exists as two isotopes.
      (iii) The isotopes occur in equal amounts.
      (iv) 80
   (b) (i) $(^{79}Br_2)^+$
      (ii) $(^{79}Br)(^{81}Br)^+$
      (iii) $(^{81}Br_2)^+$
2. $(20 \times \frac{90.48}{100}) + (21 \times \frac{0.27}{100}) + (22 \times \frac{9.25}{100})$ **(1)**
   $= 20.19$ **(1)**

4. Writing formulae and equations

1. (a) sodium nitrate
   (b) silver nitrate $AgNO_3$
      sodium carbonate Na_2CO_3
      silver carbonate Ag_2CO_3
      second product $NaNO_3$
      *(all four **2 marks**, three correct **1 mark**)*
   (c) $2AgNO_3 + Na_2CO_3 \rightarrow 2NaNO_3 + Ag_2CO_3$
2. **D**
3. (a) $C_2H_5OH + 3O_2 \rightarrow 2CO_2 + 3H_2O$ *(reactants **(1)** products **(1)** balancing **(1)**)*
   (b) $Ca(OH)_2 + CO_2 \rightarrow CaCO_3 + H_2O$ *(reactants **(1)** products **(1)** balancing **(1)**)*
   (c) $Br_2 + 2KI \rightarrow I_2 + 2KBr$ *(reactants **(1)** products **(1)** balancing **(1)**)*
   (d) $MO + 2HCl \rightarrow MCl_2 + H_2O$ *(reactants **(1)** products **(1)** balancing **(1)**)*

5. Amount of substance – the mole

1. (a) A mole is the amount of substance that has as many particles **(1)** as there are atoms in exactly 12 g of carbon-12. **(1)**
   (b) **C**

2. (a) (i) 46 g
      (ii) 1.15 g
   (b) (i) 0.05 mol
      (ii) 0.0625 mol
   (c) (i) Ratio of moles P : moles O = 0.05 : 0.0625 × 2 (as there are 2 oxygen atoms in each O_2 molecule) **(1)**
         = 0.05 : 0.125
         = 2:5; formula P_2O_5 **(1)**
      (ii) $P_4 + 5O_2 \rightarrow 2P_2O_5$ OR $4P + 5O_2 \rightarrow 2P_2O_5$
3. $M_r = \frac{4.00}{0.025} = 160$ **(1)**
   $A_r = 80$, so bromine **(1)**

6. Calculating reacting masses and gas volumes

1. (a) **C**
   (b) **D**
2. (a) 19
   (b) $M_r = 75$ **(1)**
      amount $= \frac{100}{75} = 1.33$ mol **(1)**
   (c) ratio of ammonia : hydrazine azide = 16:12 = 4:3 **(1)**
      amount in mol of ammonia $= 1.33 \times \frac{4}{3} = 1.778$ mol **(1)**
      volume of ammonia $= 1.778 \times 24 = 42.7$ dm^3 **(1)**
   (d) A large volume of gas is formed when the reaction occurs / products have low density **(1)** the reaction is very fast / explosive. **(1)**

7. Types of formulae

1. (a) **B**
   (b) The simplest whole number ratio of atoms **(1)** of each element in a compound. **(1)**
2. (a) C = 40% H = 6.67% O = 53.3% **(1)**
      $\frac{40}{12} = 3.33$ $\frac{6.67}{1} = 6.67$ $\frac{53.3}{16} = 3.33$ **(1)**
      C:H:O = 1:2:1 **(1)**
      empirical formula = CH_2O **(1)**
   (b) $CH_2O = 12 + 2 + 16 = 30$ **(1)**
      $\frac{180}{30} = 6$, molecular formula $= C_6H_{12}O_6$ **(1)**
   (c) molecular formula C_6H_{14} **(1)** empirical formula C_3H_7 **(1)**

8. Calculations involving solutions

1. (a) mass $= 20.801 - 17.756 = 3.045$ g **(1)**
      amount $= \frac{3.045}{106} = 0.0287$ mol **(1)**
      concentration $= \frac{0.0287}{0.250} = 0.115$ mol dm^{-3} **(1)**
   (b) $Na_2CO_3 + 2HCl \rightarrow 2NaCl + H_2O + CO_2$ **(1)**
      amount HCl $= \frac{50}{1000} \times 0.300 = 0.015$ mol **(1)**
      amount $Na_2CO_3 = \frac{0.015}{2} = 0.0075$ mol **(1)**
      volume $= \frac{0.0075}{0.200} = 0.0375$ dm$^3 = 37.5$ cm^3 **(1)**
2. **C**
3. amount of HCl gas $= \frac{100}{24000} = 0.004167$ mol **(1)**
   volume of solution in dm^3 = 0.5 dm^3 **(1)**
   concentration $= \frac{0.004167}{0.5} = 0.00833$ mol dm^{-3} **(1)**

9. Formulae of hydrated salts

1. (a) There are water molecules attached to the ions in the lattice / lattice contains water
   (b) (i) Otherwise some of the mass loss will be the water already in the wet crucible / the water will contribute to the mass of the empty crucible.
      (ii) To ensure that the dehydration is complete.
      (iii) $23.26 - 11.20 = 12.06$ g
      (iv) $32.24 - 23.26 = 8.98$ g
      (v) amount $MgSO_4 = \frac{12.06}{120.1} = 0.100$ mol; amount water $= \frac{8.98}{18} = 0.499$ mol **(1)**
         ratio water : $MgSO_4 = \frac{0.499}{0.100} : 1 = 4.99 : 1$ **(1)**
         formula is $MgSO_4.5H_2O$ **(1)**

10. Percentage yield and atom economy

1 (a) 100%

(b) amount HCl = $\frac{100}{24}$ = 4.167 mol **(1)**
maximum amount NH$_4$Cl = 4.167 mol **(1)**
maximum mass NH$_4$Cl = 4.167 × 53.5 = 222.9 g **(1)**
% yield = $\frac{201.6}{222.9}$ = 90.4% **(1)**

2 (a) M_r(aspirin) = 180; M_r(ethanoic acid) = 60 **(1)**
atom economy = $\frac{180}{(180 + 60)}$ **(1)**
= 75% **(1)**

(b) **B**

11. Neutralisation reactions

1 (a) In a strong acid (almost) all of the molecules dissociate (to give H$^+$ ions) **(1)** but in a weak acid only a small proportion of the molecules dissociate. **(1)**

(b) **C**

(c) (i) 2CH$_3$COOH + Na$_2$CO$_3$
$\rightarrow$ 2CH$_3$COONa + H$_2$O + CO$_2$
(reactant formulae (1), product formulae (1), balancing (1))

(ii) the white solid / sodium carbonate disappears **(1)** effervescence **(1)**

2 (a) Mix solutions in pairs **(1)**. In the pair that effervesces, the solutions are sodium carbonate and sulfuric acid **(1)**. Add each of this pair to the third solution, which is potassium hydroxide **(1)**. The one that gives a temperature rise is sulfuric acid **(1)**.

(b) Wear goggles as the concentrated acid/alkali is corrosive.

12. Acid–base titrations

1 (a) So that any solid left in the container is not included in the mass.

(b) Rinse the funnel into the volumetric flask.

(c) Shake the flask.

(d) mass = 23.105 − 20.010 = 3.095 g **(1)**
amount NaHSO$_4$ = $\frac{3.095}{120.1}$ = 0.0258 **(1)**
[NaHSO$_4$] = $\frac{0.0258}{0.250}$ = 0.103 mol dm^{-3} **(1)**

(e) (i) NaHSO$_4$ + NaOH $\rightarrow$ Na$_2$SO$_4$ + H$_2$O
(sodium sulfate product (1), rest of equation (1))

(ii) colourless to pink

13. Calculations based on titration data

1 (a) (i) 1: 18.05, 2: 19.00, 3: 18.10, 4: 18.75
(values (1), two decimal places for each (1))

(ii) $\frac{(18.05 + 18.10)}{2}$ **(1)**
= 18.08 cm^3 **(1)**

(iii) $\frac{18.08}{1000}$ × 0.110 **(1)**
= 0.00199 mol **(1)**

(b) (i) 2KOH + H$_2$SO$_4$ $\rightarrow$ K$_2$SO$_4$ + 2H$_2$O
(formulae (1), balancing (1))

(ii) A 1:2 ratio of H$_2$SO$_4$: KOH, $\frac{0.00199}{2}$ **(1)**
= 0.000995 mol **(1)**

(c) $\frac{0.000995}{0.025}$ **(1)**
= 0.0398 mol dm^{-3} **(1)**

(d) **A**

(e) Some KOH would be neutralised by carbon dioxide from the air.

14. Oxidation numbers

1 KCl: −1
KClO$_3$: +5
ClF$_3$: +3
Cl$_2$: 0

2 **C**

3 (a) (i) Fe^{2+} $\rightarrow$ Fe^{3+} + e$^-$

(ii) S$_2$O$_8^{2-}$ + 2e$^-$ $\rightarrow$ 2SO$_4^{2-}$

(b) (i) +2 to +3

(ii) +6 **(1)** sulfate(VI) **(1)**

(c) 2Fe^{2+} + S$_2$O$_8^{2-}$ $\rightarrow$ 2Fe^{3+} + 2SO$_4^{2-}$

---

4 Fluorine's electronegativity is higher **(1)**, so fluorine has a negative oxidation number / −1. **(1)**

15. Examples of redox reactions

1 (a) Fe$_2$O$_3$

(b) FeO

(c) **C**

2 (a) Potassium dichromate(VI)

(b) Cr$_2$O$_7^{2-}$ + 14H$^+$ + 6e$^-$ $\rightarrow$ 2Cr^{3+} + 7H$_2$O
(formulae (1), balancing (1), electrons (1))

(c) VO$_2^+$ +5 **(1)**
VO^{2+} +4 **(1)**
[V(H$_2$O)$_6$]$^{3+}$ +3 **(1)**
[V(H$_2$O)$_6$]$^{2+}$ +2 **(1)**

3 (a) The same element is oxidised and reduced simultaneously.

(b) copper(I) is oxidised to copper(II) **(1)** copper(I) is reduced to copper(0) **(1)**
The blue solution is copper(II) sulfate and the red-brown precipitate is copper metal. **(1)**

16. Exam skills 1

1 (a) (i) % O = 100 − 34.33 − 17.91 = 47.76% **(1)**
Na: $\frac{34.33}{23}$ = 1.493; C = $\frac{17.91}{12}$ = 1.4925; O = $\frac{47.76}{16}$ = 2.985 **(1)**
Na: C: O = 1:1:2, empirical formula = NaCO$_2$ **(1)**

(ii) NaCO$_2$ = 67, formula = Na$_2$C$_2$O$_4$

(b) (i) MnO$_4^-$ + 8H$^+$ + 5e$^-$ $\rightarrow$ Mn^{2+} + 4H$_2$O
(formulae (1), balancing (1))

(ii) MnO$_4^-$ +7
Mn^{2+} +2

(c) (i) Na$_2$C$_2$O$_4$ $\rightarrow$ Na$_2$CO$_3$ + CO
(CO (1), rest of equation (1))

(ii) Na$_2$CO$_3$ = 106, CO = 28 **(1)**
atom economy = $\frac{106}{(106 + 28)}$ × 100% **(1)** = 79% **(1)**

(iii) sodium carbonate also decomposes

17. Electron shells and orbitals

1 **C**

2 (a)

(b) In the p sub-shell there are three p-orbitals **(1)**, each of which can contain two electrons. **(1)**

(c) **C**

3 **B**

18. Electron configurations – filling the orbitals

1 (a) fluorine atom 1s^2 2s^{2}2p^5 **(1)**
chlorine atom 1s^2 2s^{2}2p^6 3s^{2}3p^5 **(1)**
bromine atom 1s^2 2s^{2}2p^6 3s^{2}3p^{6}3d^{10} 4s^{2}4p^5 **(1)**

(b) There are seven electrons in the highest shell

2 (a) Ne

(b) Al^{3+}

(c) O^{2-}

3 (a) **A**

(b) (i) [Ar] 3d^{10} 4s^2 / [Ar] 4s^2 3d^{10}

(ii) [Ar] 3d^{10}

(iii) [Ar]
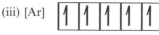

19. Ionic bonding

1 (a) (i)

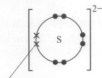

Two of the electrons are shown as crosses to show that they have been gained by the sulfur atom

(ii)

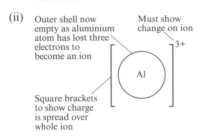

Outer shell now empty as aluminium atom has lost three electrons to become an ion

Must show change on ion

Square brackets to show charge is spread over whole ion

(b) Al_2S_3

(c) A regular arrangement (of ions) **(1)** repeated many times. **(1)**

(d) Electrostatic attraction between positive aluminium ions and negative sulfide ions **(1)** is strong / requires a large amount of energy to overcome. **(1)**

2 (a) **B**

(b) **A**

20. Covalent bonds

1 (a) The electrostatic attraction between a shared electron pair **(1)** and two nuclei. **(1)**

(b) (i) HNO_3

(ii) nitric acid

(iii) One atom donates both electrons to be shared. **(1)**

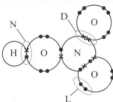

*(D circle **(1)**, N circle **(1)**, L circle **(1)**)*

(iv) A pair of electrons not used in bonding **(1)** on the outer shell. **(1)**

2

*(structure with molecular formula C_3H_6 **(1)**, all single bonds correct **(1)**, double bond **(1)**)*

21. Shapes of molecules

1 **D**

2 (a) (i)

*(five F around P with single bonds **(1)**, shape trigonal bipyramid **(1)**)*

(ii) 90° **(1)** 120° **(1)**

(iii) There are five bond pairs around P **(1)**, which repel to be as far apart as possible. **(1)**

3 (a)

*(trigonal shape **(1)**, three double bonds **(1)**)*

(b) trigonal planar **(1)**, 120° **(1)**

(c) $S_8 + 12O_2 \rightarrow 8SO_3$

22. More shapes of molecules and ions

1 (a) (i) tetrahedral

(ii) There are four bond pairs of electrons around C **(1)**, which repel to be as far apart as possible. **(1)**

(iii) 109.5°

(b) (i) pyramidal

(ii) 107° **(1)** there is a lone pair **(1)** that repels more than a bond pair. **(1)**

2 **B**

23. Electronegativity and bond polarity

1 (a) The power of an atom to attract electrons **(1)** in a covalent bond. **(1)**

(b) The electronegativity increases **(1)** because the number of protons in each atom increases **(1)** and the shielding is similar. **(1)**

(c) (i) Si/P/S–Cl (allow other feasible bonds) **(1)** δ+ on Si/P/S and δ– on Cl **(1)**

(ii) Yes **(1)**. The N–H bonds are polar **(1)** and their dipoles do not cancel. **(1)**

2 **B**

3 The carbon chain is non-polar so does not interact with water. **(1)** The O–H bond is polar (allowing H-bonds to form with polar water molecules) and causes the solubility. **(1)** The longer the carbon chain, the less the effect of the O–H bond and the less soluble is the alcohol. **(1)**

24. Van der Waals' forces

1 (a) (i) Fluctuations in the atoms' electron distribution cause a temporary dipole **(1)**, which induces a dipole in neighbouring atoms **(1)** and the attraction between dipoles is the intermolecular force. **(1)**

(ii) (atomic number on x axis and boiling point on y axis **(1)**, plotting **(1)**, line **(1)**)

(iii) As the atomic number increases the boiling point increases **(1)** because with higher atomic number there are more electrons **(1)** and increased London / dispersion forces. **(1)**

2 **C**

3 **A**

25. Hydrogen bonding and the properties of water

1 (a)

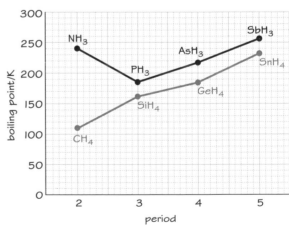

*(axes including unit **(1)**, plotting **(2)**; one mark for each Group line)*

(b) The boiling point increases as relative molecular mass increases **(1)**, the number of electrons in the molecule increases **(1)**, the London forces increase. **(1)**

(c) The difference between the electronegativities of N and H is large so the N–H bond is very polar **(1)** leading to hydrogen bonding in ammonia **(1)** and H-bonding (only found in ammonia) is stronger than other intermolecular forces (found in the other compounds). **(1)**

26. Properties of simple molecules

1 (a) Enthalpy change when one mole of a bond **(1)** in the gas phase is broken. **(1)**
   (b) The electrostatic attraction between the shared pair and nuclei is lower in I–I **(1)** because the shared pair is less attracted **(1)** as the bond is longer. **(1)**
   (c) In diamond many strong C–C bonds are broken **(1)** but in water only the weaker intermolecular forces are broken. **(1)**

2 A

3 (a) *(diagram with four or more iodine molecules **(1)** arranged regularly **(1)**)*
   (b) **B**

27. Exam skills 2

1 (a) B = 78.3; H = 100 − 78.3 = 21.7 **(1)**
      B $\frac{78.3}{10.8}$ = 7.25; H $\frac{21.7}{1}$ = 21.7 **(1)**
      B $\frac{7.25}{7.25}$ = 1; H $\frac{21.7}{7.25}$ = 2.99; B:H = 1:3; BH$_3$ **(1)**
   (b) diborane B$_2$H$_6$ **(1)**
      borane BH$_3$ **(1)**
   (c) (i)

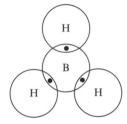

   *(three B–H bonds **(1)**, no lone pair **(1)**)*
      (ii) The three bond pairs repel to be as far apart as possible **(1)** giving a trigonal planar shape **(1)** with a bond angle of 120°. **(1)**
   (d) (i) 8BF$_3$ + 6LiH → B$_2$H$_6$ + 6LiBF$_4$
      (ii) Li +1 **(1)**
          H −1 **(1)**
      (iii) There is strong ionic attraction in LiH **(1)** but much weaker London forces in BF$_3$. **(1)**

28. The Periodic Table

1 (a) 1s^{2}2s^{2}2p^{6}3s^{2}3p^2
   (b) C
   (c) (i) The periodic table is in order of atomic number. **(1)** The atomic number / number of protons of P, 15 > Si, 14 > Al, 13 **(1)**
      (ii) Silicon has a giant structure **(1)** with strong covalent bonds joining the silicon atoms **(1)** that require a lot of energy to break. **(1)**
      (iii) Phosphorus consists of P$_4$ molecules **(1)** and only the weak intermolecular forces are overcome when phosphorus is melted. **(1)**
   (d) Silicon(IV) oxide, a non-metal oxide, is acidic **(1)**, which reacts with sodium oxide, a metal oxide, that is a base. **(1)**

29. Ionisation energy

1 (a) The energy required to remove one electron **(1)** from each of one mole **(1)** of gaseous atoms. **(1)**
   (b) (i) Mg(g) → Mg$^+$(g) + e$^-$
      *(species **(1)**, state symbols **(1)**)*
      (ii) O$^+$(g) → O^{2+}(g) + e$^-$
      *(species **(1)**, state symbols **(1)**)*

2 (a) In a potassium atom, the outer electron is further from the nucleus **(1)** and is more shielded by inner electrons **(1)** so the electrostatic attraction to the nucleus is lower. **(1)**
   (b) **C**

30. Ionisation energy across Periods 2 and 3

1 (a) D
   (b) The first ionisation energy increases across period 2 because each subsequent atom has one extra proton. **(1)** The electron being removed is in the same shell so that shielding from inner electrons is similar. **(1)** Overall, the electrostatic attraction of the outer electron to the nucleus is increased meaning that more energy is required. **(1)**
   (c) (i)

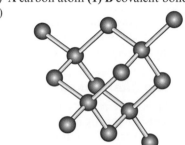

      (ii) In nitrogen there are no paired electrons in the 2p orbitals. **(1)** In oxygen, electrons are paired in one of the 2p orbitals. **(1)** These electrons repel each other. **(1)**

31. Structures of the elements

1 (a) A carbon atom **(1)** B covalent bond **(1)**
   (b)

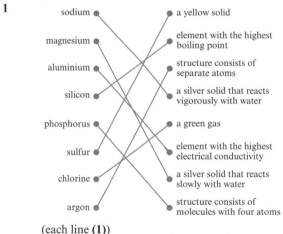

   *(each C joined to four others **(1)**, tetrahedral arrangement **(1)**)*
   (c) **C**

2 (a) Ar < Cl$_2$ < P$_4$ < S$_8$
   (b) These elements all exist as separate atoms or small molecules **(1)**. Only weak intermolecular forces are broken when the elements are boiled **(1)**, which does not require a large amount of energy **(1)**.

32. Properties of the elements

1

sodium	a yellow solid
magnesium	element with the highest boiling point
aluminium	structure consists of separate atoms
silicon	a silver solid that reacts vigorously with water
phosphorus	a green gas
sulfur	element with the highest electrical conductivity
chlorine	a silver solid that reacts slowly with water
argon	structure consists of molecules with four atoms

   (each line **(1)**)

2 Aluminium has a lattice of ions with delocalised electrons. **(1)** The delocalised electrons can move when a potential is applied. **(1)** Silicon has a limited number of electrons that can move through the structure. **(1)** Sulfur has no free electrons so is a non-conductor of electricity. **(1)**

3 (a) Mg + 2H$_2$O → Mg(OH)$_2$ + H$_2$
      *(hydrogen identified **(1)**, all other formulae **(1)**)*
   (b) Sodium hydroxide is very soluble in water. **(1)** Magnesium hydroxide is much less soluble in water. **(1)** So, sodium hydroxide solution contains a higher concentration of hydroxide ions. **(1)**

33. Group 2 elements

1 (a) Ca + 2H$_2$O → Ca(OH)$_2$ + H$_2$
      *(formulae **(1)**, balancing **(1)**)*
   (b) A white precipitate forms **(1)** and there is effervescence. **(1)**
   (c) (i) The indicator goes blue/purple **(1)** because there are hydroxide ions present. **(1)**

(ii) The solution formed in (b) is known as limewater. **(1)** Exhaled breath contains the gas carbon dioxide, which gives a white precipitate in limewater. **(1)**

2 (a) $Mg + 2HCl \rightarrow MgCl_2 + H_2$
(formulae (1), balancing (1))

(b) (i) A reaction in which electrons are transferred.

(ii) Magnesium has been oxidised **(1)** because each atom loses electrons. **(1)**

34. Group 2 compounds and their uses

1 (a) The indicator turns blue/purple **(1)** because hydroxide ions are released. **(1)**

(b) (i) Steam is given off **(1)** because the heat released is enough to boil the water. **(1)**

(ii) $SrO + H_2O \rightarrow Sr(OH)_2$
(left-hand side formulae (1), right hand side formula (1))

(c) C

2 Weigh out some tablets into a beaker **(1)**, add acid from a burette until the effervescence stops or the indicator shows neutral. **(1)** Repeat to get reliable results **(1)** from the moles of acid used, the moles and then mass of calcium carbonate can be calculated. **(1)**

35. The halogens and their uses

1 (a) Fluorine $M_r = 38$ **(1)** bromine colour = red-brown **(1)**

(b) The boiling point increases passing down the group **(1)** because the molecules have more electrons **(1)** so the intermolecular forces increase. **(1)**

2 (a) $2NaOH + Cl_2 \rightarrow NaCl + NaClO + H_2O$
(NaClO (1), everything else (1))

(b) Cl_2O **(1)** $NaCl$ −1 **(1)** $NaClO$ +1 **(1)**

(c) Chlorine as an element has an oxidation number 0 which decreases to −1 in $NaCl$ and increases to +1 in $NaClO$. **(1)** A reaction where one substance is simultaneously oxidised and reduced **(1)** is called a disproportionation reaction. **(1)**

(d) B

36. Reactivity of the halogens

1 (a) (i) $1s^2 2s^2 2p^5$

(ii) $1s^2 2s^2 2p^6$

(b) (i) $2Fe + 3F_2 \rightarrow 2FeF_3$
(formulae (1), balancing (1))

(ii) C

(iii) The halogens get less reactive passing down the group. **(1)** The atoms form negative ions in the reactions and attract the added electron more **(1)** when the atom is smaller. **(1)**

2 The brown solution when chlorine and bromine are added to iodide ion shows the formation of iodine, so iodine is the least reactive. **(1)** Chlorine displaces bromine giving an orange bromine solution **(1)**, so the order of reactivity is chlorine > bromine > iodine. **(1)**

37. Tests for ions

1 (a) $Ca(OH)_2 + CO_2 \rightarrow CaCO_3 + H_2O$
(left-hand side formulae (1), right-hand side formulae (1))

(b) (i) white precipitate / milky appearance

(ii) calcium carbonate

2 (a) A

(b) (i) 1 Add the sample to a test tube and dissolve.
2. To the solution, add dilute nitric acid **(1)** then barium nitrate solution. **(1)**
3. A white precipitate forms, indicating sulfate ions. **(1)**

(ii) $Ba^{2+} + SO_4^{2-} \rightarrow BaSO_4$
(left-hand side formulae (1), right-hand side formula (1))

(c) $(NH_4)_2SO_4$

3 D

38. Exam skills 3

1 (a) Fluorine is a gas and astatine is a solid. **(1)** Fluorine is pale yellow / colourless. **(1)** Astatine is black. **(1)**

(b) (i) Only here is the concentration of the ions high enough.

(ii) Add silver nitrate solution. **(1)** A cream precipitate forms **(1)** that will dissolve in concentrated ammonia solution but not in dilute ammonia solution. **(1)**

(iii) Seawater also contains chloride ions **(1)**, which give a precipitate with silver nitrate. **(1)**

(c) (i) $Br(g) \rightarrow Br^+(g) + e^-$
(formulae (1), state symbols (1))

(ii) There is one extra proton in krypton's atom **(1)** and similar shielding, so more energy is required to remove the electron from its attraction. **(1)**

(d) D

39. Enthalpy profile diagrams

1 (a) A activation energy **(1)**
B enthalpy change of reaction **(1)**

(b) The enthalpy of the reactants is higher than that of the products. **(1)**

(c)

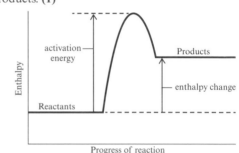

(enthalpy level of products higher than reactants (1), labelling (1))

2 (a) It is the minimum energy **(1)** that is required in order that a reaction can take place. **(1)**

(b) A

40. Enthalpy change of reaction

1 (a) This is the heat energy change **(1)** when 1 mole of solid is completely dissolved in water. **(1)**

(b) it is an insulator / does not absorb heat

(c) add a lid **(1)** to prevent heat transfer to/from the environment **(1)**

(d) take the temperature after adding the solid regularly, such as every 30s or every minute, for 10 minutes **(1)**, so that the data can be plotted on a graph that can be extrapolated to find the temperature change more accurately. **(1)**

(e) C

41. Calculating enthalpy changes

1 (a) $q = 50 \times 4.18 \times (64.2 - 18.4)$ **(1)**
$= 9572.2 J$ **(1)**
$= 9.57 kJ$ **(1)**

(b) (i) $\frac{2.45}{24.3} = 0.1008$ mol

(ii) $\frac{50}{1000} \times 2.0 = 0.10$ mol

(c) $\frac{9.57}{0.1}$ **(1)**
$= -95.7 kJ\,mol^{-1}$ **(1)**

(d) B

42. Enthalpy change of neutralisation

1 (a)

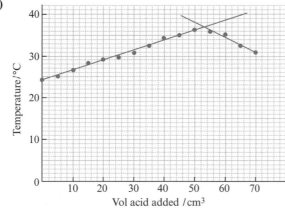

(axes and scale (1), points plotted correctly (2), two best-fit lines (1))

(b) Highest temperature 36.4 °C **(1)** Lowest temperature 24 °C **(1)** Temperature rise = 36.4 − 23.5 = 12.9° C **(1)**

(c) **C**

43. Hess' law

1 (a) $Fe_2O_3 + 3CO \rightarrow 2Fe + 3CO_2$

(b) $(-48.3) + (21.8 \times 2) + (6 \times -10.9)$ **(1)**
$= -70.1$ **(1)**
$-\frac{70.1}{3} = -23.4 \, kJ \, mol^{-1}$ **(1)**

(c) The temperature of each step included in the cycle must be the same

(d) **D**

44. Enthalpy change of formation

1 $N_2(g) + 2H_2(g) + \frac{3}{2}O_2(g) \rightarrow NH_4NO_3(s)$
(formulae (1), numbers (1), state symbols (1))

2 **D**

3 (a) $C_6H_{12}O_6(s) + 6O_2(g) \rightarrow 6CO_2(g) + 6H_2O(l)$
(formulae and balancing (1), state symbols (1))

(b) $\Delta_r H^\ominus = \Sigma \Delta_f H^\ominus \text{(products)} - \Sigma \Delta_f H^\ominus \text{(reactants)}$
$\Sigma \Delta_f H^\ominus \text{(products)} = -4080$ **(1)**
$\Sigma \Delta_f H^\ominus \text{(reactants)} = -1275$ **(1)**
$\Delta_r H^\ominus = (-4080) - (-1275) = -2805 \, kJ \, mol^{-1}$ **(1)**

(c) $\Delta_r H^\ominus = \Sigma \Delta_f H^\ominus \text{(products)} - \Sigma \Delta_f H^\ominus \text{(reactants)}$
$\Sigma \Delta_f H^\ominus \text{(products)} = -3146$ **(1)**
$\Sigma \Delta_f H^\ominus \text{(reactants)} = -2223$ **(1)**
$\Delta_r H^\ominus = (-3146) - (-2223) = -923 \, kJ \, mol^{-1}$ **(1)**

45. Enthalpy change of combustion

1 **B**

2 (a) Place some propan-1-ol in a spirit burner and weigh. **(1)** Measure some water into a container and light the burner under the water. **(1)** Take the temperature of the water at the start and the end. **(1)** Weigh the burner again to find the mass of fuel burnt. **(1)**

(b) (i) $q = mc\Delta T = 100 \times 4.18 \times 54.4 = 22739.2 \, J$
$= 22.7392 \, kJ$ **(1)**
moles of propan-1-ol = 0.900/60 = 0.015 mol **(1)**
$\Delta_c H^\ominus = -\frac{22.7392}{0.015} = -1516 \, kJ \, mol^{-1}$ **(1)**

(ii) $\frac{(2021 - 1516)}{2021} \times 100\% = 25\%$

3 $\Delta_r H^\ominus = \Sigma \Delta_c H^\ominus \text{(reactants)} - \Sigma \Delta_c H^\ominus \text{(products)}$
$\Sigma \Delta_c H^\ominus \text{(reactants)} = (6 \times -394) + (7 \times -286) = -4366$ **(1)**
$\Sigma \Delta_c H^\ominus \text{(products)} = -4163$ **(1)**
$\Delta_r H^\ominus = (-4366) - (-4163) = -203 \, kJ \, mol^{-1}$ **(1)**

46. Bond enthalpies

1 (a) $H_2(g) \rightarrow 2H(g)$

(b) The bonds in 6–9 occur in a variety of molecules, so the values are mean **(1)** but the bonds in 1–5 only occur in the diatomic molecule listed, so the values are not mean. **(1)**

(c) **C**

(d) Bond enthalpies in reactants = $(6 \times 413) + (347) + (614) + (193) = 3632 \, kJ \, mol^{-1}$ **(1)**
Bond enthalpies in products = $(6 \times 413) + (2 \times 347) + (2 \times 276) = 3724 \, kJ \, mol^{-1}$ **(1)**
$\Delta_r H^\ominus = 3632 - 3724 = -92 \, kJ \, mol^{-1}$ **(1)**

(e) $4 \times -276 = -1104 \, kJ \, mol^{-1}$

47. Collision theory

1 (a) hydrogen

(b) The gas syringe volume remains constant / effervescence stops.

(c) **C**

(d) A larger number of H^+ ions will be present per unit volume **(1)** so there is a higher rate of collisions between Zn and H^+. **(1)**

(e) The reaction begins when the reagents are mixed **(1)** so some gas will escape before the apparatus is sealed. **(1)** To avoid this, have the zinc in a tube balanced inside the sealed apparatus, and shake to tip the tube over and allow the zinc into the acid / use of thistle funnel. **(1)**

48. Measuring reaction rates

1 (a)

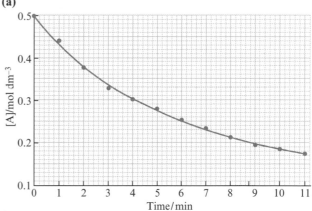

(axes labelling and scale (1), plotting points (2), best-fit curve (1))

(b)

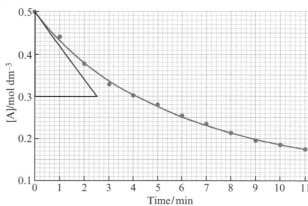

initial rate = gradient = $\frac{\Delta y}{\Delta x}$
= 0.2/ 2.5 (1, 1)
initial rate = 0.08 mol dm^{-3} min^{-1} **(1)**

(c) The graph still shows a curve / the concentration is still changing at 11 minutes **(1)** so the reaction is not complete. **(1)**

(d) **B**

49. The Boltzmann distribution

1 (a) Only a small proportion of the molecules have the activation energy. **(1)**

(b)
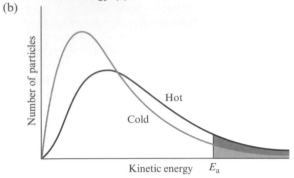

*(peak higher **(1)** and to left **(1)**)*

(c) The rate of reaction will reduce **(1)** because a smaller proportion of the molecules have energy ≥ activation energy **(1)**, so a lower proportion of the collisions will be successful. **(1)**

(d) **D**

50. Catalysts

1 (a) A measured volume of hydrogen peroxide solution was placed in a flask with a gas syringe connected and left for a period of time. The volume of any gas evolved was measured. **(1)** The experiment was repeated, but some drops of potassium iodide were added. **(1)** The experiment was repeated again, but some drops of sodium chloride were added. **(1)** Only the reaction with potassium iodide added was faster showing that iodide ions are the catalyst. **(1)**

(b) **C**

(c) A homogeneous catalyst is in the same phase as the reactants and a heterogeneous catalyst is in a different phase to at least one reactant. **(1)** In this example, the catalyst is homogeneous. **(1)**

51. Dynamic equilibrium

1 (a) The forward and backward reactions are occurring at the same rate **(1)** so that the concentrations of each substance are not altering. **(1)**

(b) (i)(ii)
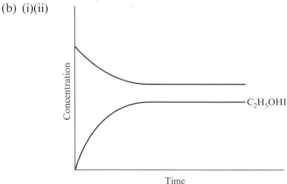

*(rising line from origin **(1)**, line rises by same amount as ethanoic acid line falls **(1)**)*

(iii) Using pipette withdraw measured sample of reaction mixture. **(1)** Add sample to large volume of iced water. **(1)** Titrate with known concentration sodium hydroxide solution. **(1)** Repeat this at regular time intervals. **(1)**

(c) **D**

(d) The concentrated sulfuric acid catalyst is corrosive, so wear gloves.

52. Le Chatelier's principle

1 Le Chatelier's principle states that when any change is made to the conditions of an equilibrium, the position of equilibrium moves **(1)** in the direction that minimises the change. **(1)**

2 (a) The position of equilibrium moves to the right **(1)** because the forward direction is endothermic. **(1)**

(b) The position of equilibrium moves to the left **(1)** because there are fewer gas molecules on the left. **(1)**

3 (a) There are more $FeSCN^{2+}$ ions **(1)** so the position of equilibrium has moved right **(1)**, so the forward reaction is exothermic. **(1)**

(b) AgSCN will form as a precipitate, lowering the concentration of SCN^- ions **(1)**, so the position of equilibrium will move to the left. **(1)**

4 **B**

53. The equilibrium constant

1 (a) $K_c = \frac{[SO_3]^2}{[SO_2]^2[O_2]}$

(b) **B**

(c) A higher temperature decreases the time required to reach equilibrium **(1)** but reduces the yield **(1)** and increases energy costs. **(1)** So, a compromise is used for a reasonable yield at a reasonable rate. **(1)**

(d) The concentrations of each gas in $mol\,dm^{-3}$ are $[SO_2]$ = $5\,mol\,dm^{-3}$, $[O_2]$ = $2.5\,mol\,dm^{-3}$, $[SO_3]$ = $15\,mol\,dm^{-3}$ **(1)** Substituting into the K_c expression, $K_c = \frac{(15)^2}{(5)^2(2.5)} = \frac{225}{62.5} = 3.6$ **(1)** Substituting $mol\,dm^{-3}$ into K_c to give units, $\frac{(mol\,dm^{-3})^2}{(mol\,dm^{-3})^2(mol\,dm^{-3})} = dm^3\,mol^{-1}$ answer = $3.6\,dm^3\,mol^{-1}$ **(1)**

2 **C**

54. Exam skills 4

1 (a) (i) $\Delta_r H^\ominus = \Sigma \Delta_f H^\ominus$ (products) $- \Sigma \Delta_f H^\ominus$ (reactants) $\Sigma \Delta_f H^\ominus$ (products) = 180.6 **(1)** $\Sigma \Delta_f H^\ominus$ (reactants) = 102.4 **(1)** $\Delta_r H^\ominus = (180.6) - (102.4) = +78.2\,kJ\,mol^{-1}$ **(1)**

(ii) Chlorine is an element

(b) Bond enthalpies in reactants = $(2 \times 481) + (2 \times 159)$ = $1280\,kJ\,mol^{-1}$ **(1)** Bond enthalpies in products = $(2 \times 481) + (243)$ = $1205\,kJ\,mol^{-1}$ **(1)** $\Delta_r H^\ominus = 1280 - 1205 = +75\,kJ\,mol^{-1}$ **(1)**

(c)

Change	Effect on rate of attainment of equilibrium	Effect on yield of chlorine
increase of temperature	increased	increased
increase of pressure	increased	decreased
addition of a catalyst	increased	no change

((1) for each row)

(d) **C**

55. Key terms in organic chemistry

1 (a) **C**

(b) An atom / group with an unpaired electron

2 (a) (i) A is and B is not **(1)** because only B has a benzene ring. **(1)**

(ii) A is and B is not **(1)** because B (only) has double bonds. **(1)**

(b) methylcyclohexane

56. Naming hydrocarbons

1 (a) 2,3-dimethylpentane **(1)** 4-methylhepta-1,6-diene **(1)** 2,3-dimethylbut-2-ene **(1)** 2,2-dimethylpropane **(1)**

(b) (i) $CH_3CH_2CH(OH)CH_3 \rightarrow CH_3CH_2CH=CH_2$ [or $CH_3CH=CHCH_3$] + H_2O
*(water **(1)**, rest of equations **(1)**)*

(ii) **B**

(iii) cyclobutane / methylcyclopropane

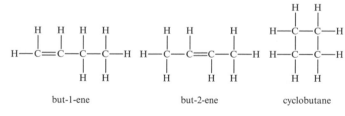

(name (1), structure (1))

57. Naming compounds with functional groups

1 (a) carboxylic acid **(1)**, alcohol / phenol **(1)**, haloalkane / chloroalkane **(1)**
   (b) $C_{11}H_7O_3Cl$
2 **D**
3 (a)

   ![displayed structure: Cl—C(Cl)(Cl)—C(H)(H)—H]

   (b) any four of:

   2-bromopropan-1-ol   3-bromopropan-1-ol   1-bromopropan-1-ol

   1-bromopropan-2-ol   2-bromopropan-2-ol

   (name (1), structure (1))

58. Different types of formulae

1 (a) **E**
   (b) **A**
   (c) **B/C**
   (d) **A and C (1)** ; **B and E (1)**
   (e) **B, D and E**
   (f) **A, B, C and E**
   (g) C_nH_{2n+2}
   (h)

   but-1-ene   but-2-ene   cyclobutane

   (allow methylcyclopropane) (name (1), structure (1))
2 (a) (i) The percentage of hydrogen $= 100 - 40 - 53.3 = 6.7\%$ **(1)**
           $\frac{40}{12} = 3.33; \frac{53.3}{16} = 3.33; \frac{6.7}{1} = 6.7$ **(1)**
           Ratio of C:O:H = 1:1:2, empirical formula COH_2 **(1)**
       (ii) $COH_2 = 30$, molecular formula $= C_2O_2H_4$
       (iii)

       ![displayed structure of ethanoic acid: H—C(H)(H)—C(=O)—O—H]

       ethanoic acid *(allow any other suitable molecule, e.g. methyl methanoate) (name (1), structure (1))*

59. Structural isomers

1 (a) Molecules with the same molecular formula **(1)** but different structural / displayed formulae. **(1)**
   (b) (i)

   ![displayed structure: H—C(H)(H)—C(H)(H)—C(H)(H)—O—H]

   (ii) propan-2-ol

   ![displayed structure of propan-2-ol]

2 (a) $C_7H_{13}Cl$
   (b) **D**
   (c)

   ![skeletal structure with double bond and terminal Cl]

   (double bond and Cl can be in any position on chain; chain can be branched)

60. Properties and reactivity of alkanes

1 (a) (i)

   ![displayed structure of methylpropane]

   (ii) **D**
   (iii) The molecule contains only carbon and hydrogen **(1)** and its formula matches C_nH_{2n+2}. **(1)**
   (iv) Methylpropane is branched so less surface area is in contact **(1)**, reducing the intermolecular forces. **(1)**
   (b) **A**

61. Reactions of alkanes

1 (a) $CH_3CH_2CH_3 + Cl_2 \rightarrow CH_3CH_2CH_2Cl + HCl$
       (structural formula of 1-chloropropane (1), rest of equation (1))
   (b) (i) Initiation: $Cl_2 \rightarrow 2Cl^{\bullet}$ **(1)**
           Propagation:
           $CH_3CH_2CH_3 + Cl^{\bullet} \rightarrow CH_3CH_2CH_2^{\bullet} + HCl$ **(1)**
           $CH_3CH_2CH_2^{\bullet} + Cl_2 \rightarrow CH_3CH_2CH_2Cl + Cl^{\bullet}$ **(1)**
           Termination: $CH_3CH_2CH_2^{\bullet} + Cl^{\bullet} \rightarrow CH_3CH_2CH_2Cl$ **(1)**
           (combination of any two radicals allowed)
       (ii) (free) radical **(1)** substitution **(1)**
   (c) breaks the bond **(1)** Cl–Cl **(1)**
   (d) **B**

62. Bonding in alkenes

1 (a) (i) In a σ bond, the shared pair of electrons **(1)** are directly between the nuclei **(1)** but in a π bond, are above and below the line joining the nuclei. **(1)**
       (ii) **D**
       (iii) **A**
   (b) (i) The molecule has (at least one) C=C double bond.
       (ii) It has a C=C bond and is hydrocarbon / formula fits C_nH_{2n}
       (iii)

       ![displayed structure with C=C double bond between 1st and 2nd carbon]

       (molecule of correct molecular formula and with one C=C (1); C=C between 1st and 2nd carbon) (1)

63. Stereoisomerism in alkenes

1 (a) Molecules with same structural formula **(1)** but where the atoms are arranged differently in space. **(1)**
  (b) (i) **A** and **D** **(1)** they have a C=C double bond **(1)** and each carbon of the double bond has two different atoms attached. **(1)**
    (ii) Z-1,2-dichloroethene *(Z **(1)**, name **(1)**)*
    cis-1,2-dichloroethene **(1)**

2 (a)

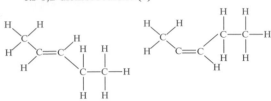

  (b)

  (c)

64. Addition reactions of alkenes

1 (a) $CH_2=CH_2 + H_2O$ **(1)** $\rightarrow C_2H_5OH$ **(1)**
  (b) steam **(1)**; catalyst of phosphoric acid **(1)**
  (c) In the addition reaction, a π bond **(1)** is broken and only one product is formed. **(1)**

2 (a)

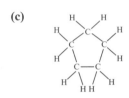

  *(δ+, δ− correct way round on Br₂ **(1)**, curly arrow from Br⁻ lone pair to carbon **(1)**, + on carbon **(1)**)*
  (b) **B**
  (c) cyclohexane – remains orange **(1)**
    cyclohexene – from orange to colourless **(1)**

65. Formation and disposal of polymers

1 (a) (i)

    (ii) The π bonds in the monomer molecules break **(1)** and the monomers bond to form a long chain. **(1)**
  (b) **D**

2 (a) poly(chloroethene)
  (b) On combustion a large volume of waste is removed **(1)** and heat is released, which is useful energy. **(1)** But HCl is released **(1)**, which must be removed because it is toxic. **(1)**

66. Exam skills 5

1 (a)

  (b) **B**
  (c)

    1,1-dibromoethane **(1)**   **(1)**

2 (a) alkenes
  (b) alcohols
  (c)

  (d)

  (e)

67. The properties of alcohols

1 (a) From methane to propane, the boiling points increase. **(1)** This is because molecules have more electrons and hence stronger London forces. **(1)**
  (b) **C**
  (c)

    propan-1-ol   propan-2-ol

  (structures one mark each; allow 1 mark if correct but no O–H bonds)
  The intermolecular forces in propan-1-ol are higher **(1)** because the contact between the molecules is greater. **(1)**
  (d) (i) methanol, ethanol and propan-1-ol
    (ii) propan-2-ol
    (iii) none

68. Combustion and oxidation of alcohols

1 (a) (i) tertiary alcohol
    (ii) $CH_3C(CH_3)(OH)CH_2CH_3$
  (b) (i) orange **(1)** to green **(1)**
    (ii) primary or secondary alcohol
    (iii) oxidising agent

2 The aldehyde distils off as soon as it is formed. **(1)** If heated under reflux it would be oxidised to a carboxylic acid. **(1)**

69. More reactions of alcohols

1 (a) **D**
  (b) phosphoric acid / sulfuric acid / ceramic **(1)** heat **(1)**
  (c) $CH_3CH_2CH_2CH=CH_2$ pent-1-ene **(1)**
    $CH_3CH_2CH=CH_2CH_3$ pent-2-ene **(1)**

2 (a)

    OH

  (b) **A**
  (c) sodium bromide **(1)**, sulfuric acid **(1)** *(allow alternative halides such as PBr₃)*

70. Nucleophilic substitution reactions of haloalkanes

1 (a) 1,2-dibromopropane
 (b)

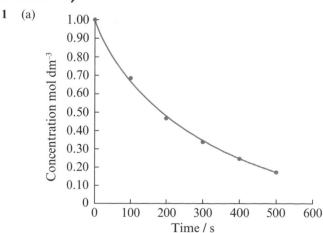

 (c) **B**
 (d) step 1 bromine
  step 2 sodium hydroxide
2 (a) silver iodide **(1)**; yellow **(1)**
 (b) All substances dissolve in the ethanol and mix / to dissolve the haloalkane.
 (c) The C–I bond is the weakest, then C–Br then C–Cl, so the C–I bond breaks most easily and reacts fastest.

71. Preparing a liquid haloalkane

1 (a) The alcohol is flammable, so naked flames are avoided.
 (b) Condenser **(1)** to condense vapours and stop them escaping from the apparatus, which would lose reactants or products / release flammable or harmful substances. **(1)**
 (c) Lower opening **(1)** to avoid air pockets in the condenser. **(1)**
 (d) (i) This is an alkali so it will neutralise the acid.
  (ii) anhydrous calcium chloride / sodium sulfate / magnesium sulfate/calcium sulfate

72. Haloalkanes in the environment

1 (a) 1,1,1-trichloro-2-fluoroethane
 (b) $C_2H_2Cl_3F$
 (c) (i) An atom or group of atoms with an unpaired electron.
  (ii) $Cl^•$
  (iii) **B**
 (d) (i) $C_2H_2Cl_3F \rightarrow {}^•C_2H_2Cl_2F + Cl^•$ **(1)**
   $Cl^• + O_3 \rightarrow ClO^• + O_2$ **(1)**
   $ClO^• + O_3 \rightarrow Cl^• + 2O_2$ **(1)**
   overall $2O_3 \rightarrow 3O_2$ **(1)**
  (ii) Ozone protects Earth from UV light **(1)**, which would cause cancer. **(1)**

73. Organic synthesis

1 (a) (i) UV radiation
  (ii) because chlorine forms free radicals **(1)** and chlorine replaces hydrogen **(1)**
  (iii) $CH_3CH_2CH_2CH_2Cl$ **(1)**
   $CH_3CH_2CHClCH_3$ **(1)**
  (iv) $CH_3CH_2CH_2CH_3 + Cl_2 \rightarrow CH_3CH_2CH_2CH_2Cl$ / $CH_3CHClCH_2CH_3 + HCl$

   (*HCl* **(1)**, *rest of equation* **(1)**)
 (b) 2-chlorobutane
 (c) (i) steam / water
  (ii) **A**
  (iii) Markownikoff's rule predicts that butan-2-ol is the major product. **(1)**
   This forms via a secondary carbocation which is more stable. **(1)**

74. Exam skills 6

1 (a) From orange/yellow/red to colourless.
 (b) (i) **D**
  (ii) drying agent: anhydrous calcium chloride / calcium sulfate / sodium sulfate
   observation: mixture goes clear
   method: filtration
  (iii) distillation
 (c) (i)

 (ii) 1-Bromohex-1-ene **(1)** because the end carbon has two hydrogen atoms. **(1)**

75. Infrared spectroscopy

1 (a) **C**
 (b) (i) Any two from: CH_3CH_2OH **(1)**, CH_3CHO **(1)**, CH_3OCH_3 **(1)**
  (ii) There is a peak at $1725\,cm^{-1}$ (allow 1710–1740), which shows a C=O bond **(1)**, but no broad peak at 3200–$3600\,cm^{-1}$, which shows no O–H bond **(1)** so the substance is ethanal. **(1)**

76. Uses of infrared spectroscopy

1 (a) **B**
 (b) When exposed to IR radiation, bonds in the molecules vibrate more **(1)**. This prevents the radiation passing into space **(1)** and some of the energy subsequently goes back to Earth. **(1)**
 (c) (i) (acidified) potassium dichromate(VI)
  (ii) **A** is ethanoic acid **(1)** **B** is ethanal **(1)** *(allow one mark if wrong way round)*
   A: because there is a broad peak at $3050\,cm^{-1}$ showing an O–H group **(1)**
   B: because there is a peak at $1720\,cm^{-1}$ showing a C=O group but no broad peak at 2500–$3300\,cm^{-1}$ showing an O–H group **(1)**

77. Mass spectrometry

1 (a) (i) *m*: mass
   z: charge
  (ii) molecular ion
 (b) (i) 46
  (ii) caused by molecules with an isotope (of C/H) **(1)** with atomic mass that is heavier **(1)**
 (c) (i) $CH_3CH_2^+$
  (ii) CH_2OH^+
  (iii) ethanol / CH_3CH_2OH

78. Concentration–time graphs (zero order reactants)

1 (a) It is a catalyst.
 (b) measuring cylinder
 (c) The sodium hydrogencarbonate reacts with the sulfuric acid **(1)** so the reaction slows down. This is called quenching **(1)**.
 (d) **C**

79. Concentration–time graphs (first order reactants)

1 (a)

 (axes: **(1)**, plotting points: **(2)**, lines of best-fit: **(1)**)
 (b) (i) time from 0s to 185s half life = 185s
  (ii) time from 200s to 405s half life = 205s
 (c) The half-lives are approximately equal **(1)** therefore the decomposition is first order with respect to X **(1)**.
2 **C**

80. Rate equation and rate constant

1 (a) NO: 2 Cl_2: 1 (1) overall: 3 (1)

(b) (i) $k = \dfrac{0.010}{(0.13)^2(0.20)}$ (1)

$= 3.0$ (1)

units $= \dfrac{mol\,dm^{-3}\,s^{-1}}{(mol\,dm^{-3})^2(mol\,dm^{-3})} = dm^6\,mol^{-2}\,s^{-1}$ (1)

(ii) rate $= 3.0(2.00)^2(2.00)$ (1)

$= 24\,mol\,dm^{-3}\,s^{-1}$ (1)

(c) At the start the concentrations of the reactants are those given (1) but as the reaction proceeds the concentrations alter and the rate alters. (1)

2 B

81. Finding the order

1 (a) D

(b) Large excess means that there is much more of that substance present than is required for a complete reaction (1), which means that the concentration of this substance does not fall significantly during the reaction (1), which means that the rate of reaction is not significantly changed as this reactant's concentration varies. (1)

2 A: first order (1) B: second order (1) C: second order (1) overall order: 5 (1)

82. The rate-determining step

1 (a) $(CH_3)_3CBr + NaOH \rightarrow (CH_3)_3COH + NaBr$ (1)

(b) (i) rate $= k[(CH_3)_3CBr])$(1)

(ii) Step 1 is slower because a C–Br covalent bond has to be broken (1) but in Step 2 the oppositely charged particles attract and a bond is formed (1).

(c) (i) It is a one step reaction/ the slow step has one CH_3CH_2Br (1) molecule reacting with a hydroxide ion. (1)

(ii)

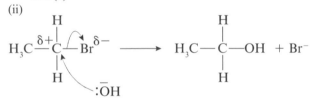

(dipole on C–Br bond : (1), curly arrow from hydroxide: (1), curly arrow to Br : (1))

(iii) C

83. The Arrhenius equation

1 (a) k: rate constant, A: pre-exponential factor, E_A: activation energy, R: Gas constant, T: temperature *(5 correct: (3), 4 correct: (2), 3 correct: (1))*

(b) (i)

283	3.52×10^{-7}	−14.9	556	1.80×10^{-3}
356	3.02×10^{-5}	−10.4	629	1.58×10^{-3}
393	2.19×10^{-4}	−8.43	666	1.50×10^{-3}
427	1.16×10^{-3}	−6.76	700	1.43×10^{-3}
508	3.95×10^{-2}	−3.23	781	1.28×10^{-3}

(each correct column: (1))

(ii)

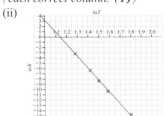

(scale: (1), plotting: (1), best-fit line: (1))

(iii) intercept on y axis $= 3$ (1)

$A = e^3$ (1)

$=$ (1)

(iv) gradient $= 1.8/18 = 0.1$ (1)

gradient $= -E_a/R$ (1)

$E_a = -\,-0.1 \times 8.31 \times 8.31$ (1)

$= 0.831\,kJ\,mol^{-1}$ (1)

84. Exam skills 7

1 (a)

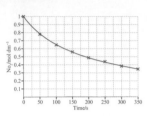

(axes and scale: (1), plotting points: (1), best-fit line: (1))

(b) C

2 Experiments 1 and 2: $[H_2] \times 1.5$, $[NO] \times 1$, rate $\times 1.5$ (1) order w.r.t. $H_2 = 1$ (1)

Experiments 3 and 2: $[H_2] \times \frac{1}{2}$ which will give rate $\times \frac{1}{2}$ (1)

$[NO] \times 6$ and rate $\times 18$, effect of NO is rate $\times 36$, order w.r.t. NO $= 2$ (1)

85. Finding the equilibrium constant

1 (a) The mixture becomes less purple. (1)

(b) $K_c = \dfrac{[HI]^2}{[H_2][I_2]}$ (1)

(c) The moles of hydrogen required to form 0.88 mol HI $= 0.44$ mol (1)

The moles of hydrogen remaining $= 1.00 - 0.44$ $= 0.56$ mol

Moles of iodine $= 0.56$ mol (1)

(d) $K_c = \dfrac{\left(\frac{0.88}{V}\right)^2}{\left(\frac{0.56}{V}\right)\left(\frac{0.56}{V}\right)}$ (1)

$= 2.47$ (1)

(e) D

(f) There are two moles of gas on each side of the equilibrium (1) so the volumes cancel (1).

(g) at equilibrium, amount $H_2 = 2 - \dfrac{0.75}{2} = 1.625$ mol (1)

amount $I_2 = 1 - \dfrac{0.75}{2} = 0.625$ mol (1)

$K_c = \dfrac{\left(\frac{0.75}{V}\right)^2}{\left(\frac{1.625}{V}\right)\left(\frac{0.625}{V}\right)}$ (1)

$= 0.55$ (1)

86. Calculating the equilibrium constant, K_c

1 In a homogeneous equilibrium, all of the reactants and products are in the same phase (1) but in a heterogeneous equilibrium the reactants and products are in at least two different phases (1).

2 (a) $K_c = \dfrac{[NH_3]^2}{[N_2][H_2]^3}$

(b) $K_c = [CO_2]$

(c) D

3 mol NaOH $= \dfrac{16.7}{1000} \times 1.0 = 0.0167$ mol (1)

mol ethanoic acid at equilibrium $= 0.0167$ mol (1)

mol ethanoic acid at start $= \dfrac{3.0}{60} = 0.05$ mol

mol ethanoic acid reacted in forming equilibrium $= 0.05 - 0.0167 = 0.0333$ mol (1)

mol ethanol at start $= \dfrac{2.3}{46} = 0.05$ mol

mol ethanol at equilibrium $= 0.05 - 0.0333 = 0.0167$ mol (1)

$K_c = \dfrac{\left(\frac{0.0333}{V}\right)\left(\frac{0.0333}{V}\right)}{\left(\frac{0.0167}{V}\right)\left(\frac{0.0167}{V}\right)}$ (1)

$= 4.0$ (1)

87. Calculating K_p

1 (a) nitrogen $= \frac{0.5}{0.8} = \frac{5}{8}$ or 0.625
    oxygen $= \frac{0.1}{0.8} = \frac{1}{8}$ or 0.125
    argon $= \frac{0.2}{0.8} = \frac{1}{4}$ or 0.25
    (3 correct: (2), 2 correct: (1))
  (b) nitrogen $= 0.625 \times 4 = 2.5$ atm, oxygen $= 0.125 \times 4$
    $= 0.5$ atm, argon $= 0.25 \times 4 = 1$ atm **(1)**

2 (a) To form 0.10 mol of S, $\frac{0.10}{2} = 0.05$ mol of R are required.
    This leaves $0.40 - 0.05 = 0.35$ mol of R at equilibrium. **(1)**
  (b) R $= \frac{0.35}{0.45} = 0.778$ **(1)**
    S $= \frac{0.10}{0.45} = 0.222$ **(1)**
  (c) R $= 0.778 \times 1000 = 778$ Pa **(1)**
    S $= 0.222 \times 1000 = 222$ Pa **(1)**
  (d) $K_p = \frac{p(S)^2}{p(R)}$ **(1)**
    $= \frac{222^2}{778}$ **(1)**
    $= 63.3$ Pa **(1)**
  (e) **B (1)**

88. The equilibrium constant under different conditions

1 **C**
2 (a) (i) $K_c = \frac{[SO_3]^2}{[SO_2]^2[O_2]}$ **(1)**
    $= \frac{\left(\frac{7.5}{2}\right)^2}{\left(\frac{2}{2}\right)^2\left(\frac{0.8}{2}\right)}$ **(1)**
    $= 35.2$ mol^{-1} dm^3 **(1)**
    (ii) The gases have not been left for enough time so
    equilibrium has not been reached/ the concentration
    of SO_3 has not reached equilibrium value. **(1)**
  (b) The position of equilibrium moves to the left **(1)** because
    there are more gas moles on the left **(1)** in order that K_c is
    kept constant. **(1)**

89. Brønsted Lowry acids and bases

1 (a) a proton donor **(1)**
  (b) **B**
2 base NH_3 **(1)** conjugate acid NH_4^+ **(1)**
3 (a) (i) $NaOH(aq) + HCl(aq) \rightarrow NaCl(aq) + H_2O(l)$
    *(left-hand side: (1), right-hand side: (1),
    state symbols: (1))*
    (ii) $OH^-(aq) + H^+(aq) \rightarrow H_2O(l)$ *(formulae: (1),
    state symbols: (1))*
    (iii) ions that remain unchanged in the reaction mixture
    (1) Na^+ and Cl^- **(1)**
  (b) (i) $HNO_3(aq) + KOH(aq) \rightarrow KNO_3(aq) + H_2O(l)$
    *(left-hand side: (1), right-hand side: (1),
    state symbols: (1))*
    (ii) $OH^-(aq) + H^+(aq) \rightarrow H_2O(l)$ *(formulae: (1),
    state symbols: (1))*
    (iii) the same reaction is occurring

90. pH

1 (a) $pH = -\log[H^+]$
  (b) (i) one where almost all of the molecules are dissociated
    giving H^+ ions
    (ii) $-\log(0.01) = 2$
    (iii) equation is $HCl + NaOH \rightarrow NaCl + H_2O$ so a
    $1:1$ ratio
    mol of HCl $= \frac{25}{1000} \times 0.01 = 0.00025$ mol
    mol of NaOH $= \frac{20}{1000} \times 0.005 = 0.0001$ mol **(1)**
    after mixing, mol of $H^+ = 0.00025 - 0.0001 =$
    0.00015 mol in 45 cm^3 mixture **(1)**
    $[H^+] = \frac{0.00015}{0.045} = 0.00333$ mol dm^{-3} **(1)**
    $pH = -\log(0.00333) = 2.48$ **(1)**

2 (a) Each molecule releases one H^+ ion.
  (b) **D**
3 $[H^+]$ in pH 1 $= 10^{-1.0} = 0.1$ mol dm^3 **(1)**
    mol H^+ in 100 cm$^3 = \frac{100}{1000} \times 0.1 = 0.01$ mol **(1)**
    $[H^+]$ in pH 2 $= 10^{-2.0} = 0.01$ mol dm^{-3} **(1)**
    To make 0.01 mol into a 0.01 mol dm^{-3} solution, need
    1000 cm^3, so add 900 cm^3 **(1)**.

91. The ionic product of water

1 (a) $2H_2O \rightleftharpoons H_3O^+ + OH^-$ / $H_2O \rightleftharpoons H^+ + OH^-$
  (b) $K_W = [H^+][OH^-]$
  (c) (i) $[H^+] = \sqrt{(4.5 \times 10^{-15})}$ **(1)**
    $= 6.7 \times 10^{-8}$ mol dm^{-3} **(1)**
    (ii) $-\log(6.7 \times 10^{-7}) = 7.2$
    (iii) **B**

2 $[OH^-] = 0.001$ mol dm^{-3}
    $K_W = [H^+][OH^-] = 4.5 \times 10^{-15}$ mol^2 dm^{-6}
    $[H^+] = 4.5 \times 10^{-15} / 0.001 = 4.5 \times 10^{-10}$ **(1)**
    $pH = -\log(4.5 \times 10^{-10}) = 11.3$ **(1)**

3 amount $OH^- = \frac{100}{1000} \times 0.015 = 0.0015$ mol
    amount $H^+ = \frac{75}{1000} \times 0.0080 = 0.0006$ mol
    excess $OH^- = 0.0015 - 0.0006 = 0.0009$ mol **(1)**
    $[OH^-] = \frac{0.0009}{0.175} = 5.14 \times 10^{-3}$ mol dm^{-3} **(1)**
    $[H^+] = \frac{1.0 \times 10^{-14}}{5.14} \times 10^{-3} = 1.94 \times 10^{-12}$ mol dm^{-3} **(1)**
    $pH = -\log(1.94 \times 10^{-12}) = 11.7$ **(1)**

92. The acid dissociation constant

1 (a) (i) $CH_3COOH + H_2O \rightleftharpoons CH_3COO^- + H_3O^+$
    $CH_3COOH \rightleftharpoons CH_3COO^- + H^+$
    (ii) $K_a = \frac{[CH_3COO^-][H_3O^+]}{[CH_3COOH]}$
    (iii) $10^{-4.75}$ **(1)**
    $= 1.78 \times 10^{-5}$ mol dm^{-3} **(1)**
  (b) (i)

    (ii) Chloroethanoic acid has a lower pK_a value which
    shows that the dissociation equilibrium is further
    to the right **(1)** giving a higher $[H^+]$ and therefore a
    stronger acid. **(1)**
    This is because chlorine has a high electronegativity,
    so weakening the O–H bond in the acid group. **(1)**

2 (a) $K_a = \frac{[H^+]^2}{[HCOOH]}$ **(1)**
    $[H^+] = \sqrt{(1.8 \times 10^{-4} \times 0.20)} = 0.006$ mol dm^{-3} **(1)**
    $pH = -\log(0.006) = 2.22$ **(1)**
  (b) **B**

93. Approximations made in weak acid pH calculations

1 (a) $C_6H_5OH + NaOH \rightarrow C_6H_5ONa + H_2O$
  (b) $K_a = \frac{[H^+]^2}{[C_6H_5OH]}$ **(1)**
    $[H^+] = \sqrt{(1.0 \times 10^{-10} \times 0.25)} = 5 \times 10^{-6}$ mol dm^{-3} **(1)**
    $pH = -\log(5 \times 10^{-6}) = 5.3$ **(1)**
  (c) The concentration of phenol is the same as its
    undissociated concentration. **(1)**
    The dissociation of water is negligible. **(1)**
  (d) The phenol value will be more accurate **(1)** because the
    assumptions are less accurate for acids that are stronger
    (1) so for trichloroethanoic acid, its concentration will be
    significantly less than its undissociated value **(1)**.

2 (a) **D**
  (b) conjugate base

94. Buffers

1 (a) A mixture that maintains an almost constant pH **(1)** when small amounts of acid or alkali are added **(1)**.
   (b) **B**
2 (a) **D**
   (b) Outside the pH range acidosis or alkalosis can occur/ enzymes are denatured **(1)**, which can lead to death. **(1)**
   (c) (i) carbonic acid/ H_2CO_3 **(1)**
       (ii) hydrogencarbonate ions/ HCO_3^- **(1)**
3 (a) $CH_3CH_2COOH(aq) + H_2O(l) \rightleftharpoons CH_3CH_2COO^-(aq) + H_3O^+(aq)$ **(1)**
       $CH_3CH_2COOK(aq) \rightarrow CH_3CH_2COO^-(aq) + K^+(aq)$ **(1)**
   (b) The added H^+ ions are removed **(1)** by reacting with propanoate ions. **(1)**

95. Buffer calculations

1 mol propanoic acid $= \frac{150}{1000} \times 0.100 = 0.015\,mol$
   $[CH_3CH_2COOH]$ in mixture $= \frac{0.015}{0.225} = 0.06667\,mol\,dm^{-3}$ **(1)**
   mol sodium propanoate $= \frac{75}{1000} \times 0.150 = 0.01125\,mol$
   $[CH_3CH_2COONa]$ in mixture $= \frac{0.01125}{0.225} = 0.05\,mol\,dm^{-3}$ **(1)**
   $K_a = [H^+](0.05)/\,0.06667$
   $[H^+] = 1.3 \times 10^{-5} \times \frac{0.06667}{0.05} = 1.73 \times 10^{-5}\,mol\,dm^{-3}$ **(1)**
   pH $= 4.76$ **(1)**
2 **D**
3 (a) $[CH_3COO^-] = \frac{150}{400} \times 0.100 = 0.0375\,mol\,dm^{-3}$ **(1)**
       $[CH_3COOH] = \frac{250}{400} \times 0.200 = 0.125\,mol\,dm^{-3}$ **(1)**
       $[H^+] = 1.8 \times 10^{-5} \times \frac{0.125}{0.0375} = 6 \times 10^{-5}\,mol\,dm^{-3}$ **(1)**
       pH $= -\log(6 \times 10^{-5}) = 4.22$ **(1)**
   (b) amount $CH_3COO^- = (\frac{150}{1000} \times 0.100) + (\frac{50}{1000} \times 0.100)$
       $= 0.02\,mol$ **(1)**
       amount $CH_3COOH = (\frac{250}{1000} \times 0.200) - (\frac{50}{1000} \times 0.100)$
       $= 0.045\,mol$ **(1)**
       $[H^+] = 1.8 \times 10^{-5} \times (\frac{0.045}{0.450})\,/\,(\frac{0.02}{0.450}) = 4.05 \times 10^{-5}\,mol\,dm^{-3}$ **(1)**
       pH $= -\log(4.05 \times 10^{-5}) = 4.39$ **(1)**

96. pH titration curves

1 (a) **B**
   (b) pH $= 9$
   (c) In this part, the sodium hydroxide solution will have neutralised some of the ethanoic acid, so the substances present are sodium ethanoate, ethanoic acid and water. **(1)** This mixture of weak acid and salt is a buffer. **(1)**

2
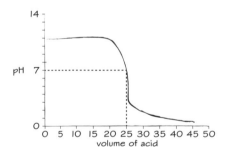

*(axes: **(1)**, shape: **(1)**, vertical region at least 4–7: **(1)***
start pH $= -\log(10-12)$ **(1)**
end pH $= -\log(25/75 \times 1) = 0.48$ **(1)**
volume $25\,cm^3$ **(1)**

97. Indicators

1 (a) Measure the pH of two solutions of known pH **(1)**, adjust the reading on the meter to match to the known value **(1)**.
   (b) (i) Cresol red and phenolphthalein **(1)**. (The vertical part of the pH graph) is from about 7–10 within which is the pH range of these indicators **(1)**.
       (ii) A weak acid–weak base titration has a pH curve with no vertical region **(1)** so all of the indicators would have a gradual colour change **(1)**.
2 left: colourless **(1)**, right: pink **(1)**

98. Exam skills 8

1 (a) (i) The yield of sulfur trioxide increases **(1)** because the position of equilibrium moves to the side with fewer gas molecules, which is right. **(1)**
       (ii) The yield of sulfur trioxide decreases **(1)**, because the position of equilibrium moves in the endothermic direction, which is left. **(1)**
   (b) amounts at equilibrium: $SO_2 = 0.55\,mol$ **(1)**,
       $O_2 = 0.275\,mol$ **(1)**, $SO_3 = 1.45\,mol$
       $$K_p = \frac{\left(\frac{1.45}{2.275} \times 35\right)^2}{\left(\frac{0.55}{2.275} \times 35\right)^2 \left(\frac{0.275}{2.275} \times 35\right)}$$ **(1)**
       $= 1.64\,kPa^{-1}$ **(1)**
   (c) (i) Each acid molecule can release two H^+ ions. **(1)**
       (ii) amount $H^+ = 2 \times \frac{25}{1000} \times 0.100 = 0.005\,mol$
       amount $OH^- = \frac{75}{1000} \times 0.100$ **(1)** $= 0.0075\,mol$ **(1)**
       pH $= -\log(1 \times \frac{10^{-14}}{(0.0075 - 0.005)}) = 11.4$ **(1)**
       (iii)

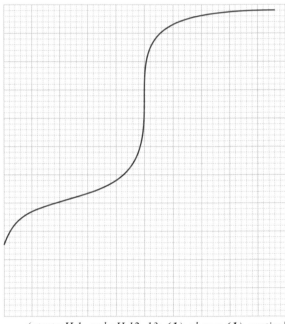

*(start pH 1, end pH 12–13: **(1)**, shape: **(1)**, vertical region at least 4–10: **(1)**)*

99. The Born–Haber cycle

1 (a) **A**
   (b) (i) Step 1: enthalpy of formation of lithium fluoride **(1)**
       (ii) Step 3: first ionisation energy of lithium **(1)**
       (iii) Step 5: first electron affinity of fluorine **(1)**
   (c) lattice enthalpy
       = step 6 = – step 5 – step 4 – step 3 – step 2 + step 1
       $= (+328) - (+79) - (+520) - (+159) + (-616)$ **(1)**
       $= -1046\,kJ\,mol^{-1}$ **(1)**

100. Factors affecting lattice enthalpy

1 (a) the enthalpy change when one mole of lithium fluoride **(1)** is formed from its gaseous ions **(1)**
   (b) Moving down Group 7, the size of the halide ions gets larger **(1)** so the electrostatic attraction to the lithium ion in the lattice gets smaller **(1)** and the lattice enthalpy becomes less exothermic. **(1)**

(c) Moving down Group 1, the size of the metal ions gets larger **(1)** so the electrostatic attraction to the fluoride ion in the lattice gets smaller **(1)** and the lattice enthalpy becomes less exothermic. **(1)**

(d) (i) Na^+ $1s^2 2s^2 2p^6$ **(1)**, Mg^{2+} $1s^2 2s^2 2p^6$ **(1)**
   (ii) MgO would be more exothermic **(1)** because the magnesium ions and oxide ions have a higher charge **(1)** and magnesium ions are smaller than the sodium ions. **(1)**

101. The enthalpy change of solution

1 (a) (i) $Ca^{2+}(g) + 2Cl^-(g) \rightarrow CaCl_2(s)$ *(equation: (1), state symbols: (1))*
      (ii) $Ca^{2+}(g) (+ aq) \rightarrow Ca^{2+}(aq)$ *(equation: (1), state symbols: (1))*
      (iii) $CaCl_2(s) (+ aq) \rightarrow Ca^{2+}(aq) + 2Cl^-(aq)$ *(equation: (1), state symbols: (1))*

   (b) (i)
   $CaCl_2(s) \longrightarrow Ca^{2+}(aq) + 2Cl(aq)$
   $\searrow \quad \nearrow$
   $Ca^{2+}(g) + 2Cl^-(g)$

   (cycle: (1), all species: (1))
      (ii) $\Delta_{sol}H = -(-2255) + (-1650) + (2 \times -338)$ **(1)**
         $= -71\,kJ\,mol^{-1}$ **(1)**

   (b) A

102. Entropy

1 B **(1)**
2 (a) $Fe_2O_3 + 3CO \rightarrow 2Fe + 3CO_2$
   (b) $\Delta_r H^\ominus = \Delta_f H^\ominus$ (products) $- \Delta_f H^\ominus$ (reactants)
      $= (3 \times -394) - (-824 + (3 \times -111))$ **(1)**
      $= (-1182) - (-1157)$ **(1)**
      $= -25\,kJ\,mol^{-1}$ **(1)**
   (c) S(products) $= 2 \times 27.3 + 3 \times 213.6 = 695.4\,J\,K^{-1}\,mol^{-1}$ **(1)**
      S(reactants) $= 87.4 + 3 \times 197.6 = 680.2\,J\,K^{-1}\,mol^{-1}$ **(1)**
      $\Delta S^\ominus = 695.4 - 680.2 = 15.2\,J\,K^{-1}\,mol^{-1}$ **(1)**
   (d) It is positive showing a small increase in disorder **(1)** because the disorder in the iron lattice and carbon dioxide molecules is more than that in the reactants **(1)**.
   (e) Because the particles in the liquid are moving around but they are ordered into a lattice in the solid.

103. Free energy

1 (a) $\Delta G = \Delta H - T \Delta S$
   (b) (i) The reaction may occur, thermodynamically/ where ΔG is negative.
      (ii) B
2 (a) $\Delta_r H^\ominus = (-110.5) - (-74.9 + -241.8)$ **(1)**
      $= +206.2\,kJ\,mol^{-1}$ **(1)**
   (b) $\Delta S^\ominus = (198 + 3 \times 131) - (186 + 189) = +216\,J\,K^{-1}\,mol^{-1}$**(1)**
   (c) $\Delta G = 206.2 - 398 \times 0.216$ **(1)**
      $= +120.2\,kJ\,mol^{-1}$ **(1)**
   (d) $\Delta G = 0$ when $\Delta H = T\Delta S$ **(1)**
      $T = \frac{206.2}{0.216} = 955\,K$ **(1)**

104. Redox

1 (a) (i) 0 **(1)**
      (ii) +5 **(1)**
   (b) reduced species: peroxodisulfate ions **(1)**, reducing agent: bromine **(1)**
2 (a) when a species loses electrons
   (b) $Cr^{3+} + 8OH^- \rightarrow CrO_4^{2-} + 4H_2O + 3e^-$ *(species: (1), balancing: (1))*
   (c) (i) $Cr_2O_7^{2-}$ +6 **(1)**, Cr^{3+} +3 **(1)**
      (ii) orange to green **(1)**
      (iii) $CH_3CH_2OH + H_2O \rightarrow CH_3COOH + 4H^+ + 4e^-$ *(formulae: (1), balancing: (1))*
      (iv) $2Cr_2O_7^{2-} + 16H^+ + 3CH_3CH_2OH$
               $\rightarrow 4Cr^{3+} + 11H_2O + 3CH_3COOH$
      (formulae: (1), balancing: (1))

105. Redox titrations

1 (a) So that the iron is gradually released in the stomach when stomach acid breaks down the coating.
   (b) It is in excess.
   (c) The residue should be washed with some of the distilled water used to make the solution up to $100\,cm^3$.
   (d) C
   (e) At the end point the purple colour of the unreacted potassium manganate(VII) is seen so no indicator is required.
   (f) amount $MnO_4^- = \frac{24.25}{1000} \times 0.0200 = 0.000485\,mol$ **(1)**
      amount Fe^{2+} in $25\,cm^3 = 5 \times 0.000485 = 0.002425\,mol$ **(1)**
      amount Fe^{2+} in $100\,cm^3 = 0.002425 \times 4 = 0.0097\,mol$ **(1)**
      mass of $Fe^{2+} = 0.0097 \times 55.8 = 0.54126\,g$ **(1)**
      % by mass $= \frac{0.54126}{1.936} \times 100\% = 28\%$ **(1)**

106. Electrochemical cells

1 (a)

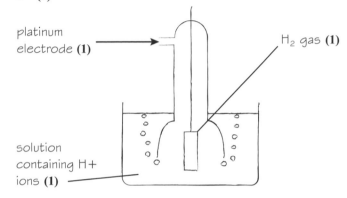

platinum electrode **(1)**

H_2 gas **(1)**

solution containing H+ ions **(1)**

   (b) Hydrogen pressure is 10^5 Pa **(1)**
      $[H^+] = 1\,mol\,dm^{-3}$ **(1)**
      temperature $= 298\,K$ **(1)**
   (c) The SHE is connected to the half-cell (with standard conditions) **(1)**, the standard electrode potential is read off from a voltmeter. **(1)**
2 (a) zinc sulfate/ zinc nitrate **(1)**, at $1\,mol\,dm^{-3}$ **(1)**
   (b) filter paper soaked in KNO_3 solution
   (c) to complete the circuit
   (d) left to right
   (e) the copper electrode/ cathode/ positive electrode

107. Measuring and using standard electrode potentials

1 (a) A beaker containing a mixture of NO_3^- and H^+ ions at each $1\,mol\,dm^{-3}$ **(1)**, and HNO_2 at $1\,mol\,dm^{-3}$ **(1)** at 298 K and with platinum electrode. **(1)**
   (b) A
   (c) B
2 (a) $2Fe^{3+} + 2I^- \rightarrow 2Fe^{2+} + I_2$ **(1)**
      $E^\ominus = +0.77 - 0.54$ **(1)**
         $= +0.23\,V$ **(1)**
      $Fe^{2+} + 2I^- \rightarrow Fe + I_2$ **(1)**
      $E^\ominus = -0.44 - 0.54$ **(1)**
         $= -0.98\,V$ **(1)**
   (b) The answer to part (a) shows that the positive $E^\ominus$ value is only for the first equation **(1)** so the final product will be Fe^{2+}. **(1)**

108. Predicting feasibility

1 (a) $2MnO_4^- + 6H^+ + 5H_2O_2 \rightarrow 2Mn^{2+} + 8H_2O + 5O_2$
   (b) $E^\ominus = +1.52 - 0.68 = +0.84\,V$
   (c) It is feasible **(1)** because $E^\ominus$ is positive. **(1)**
   (d) The solution goes from purple to colourless **(1)** and effervescence is seen. **(1)**
2 C

3 $2V^{2+} + O_2 + 2H^+ \rightarrow 2V^{3+} + H_2O_2$ **(1)**
$E^\ominus = +0.26 + 0.68 = +0.94\,V$ **(1)**
$2V^{3+} + 2H_2O + O_2 \rightarrow 2VO^{2+} + 2H^+ + H_2O_2$ **(1)**
$E^\ominus = -0.34 + 0.68 = +0.34\,V$ **(1)**
The final species is VO^{2+}, which will not be oxidised to VO_2^+ as the potential for this change would be negative **(1)**.

109. Storage and fuel cells

1 (a) anode: negative terminal **(1)**
cathode: MnO_2/ positive terminal **(1)**
(b) (i) $Zn + 2MnO_2 \rightarrow Mn_2O_3 + ZnO$
(ii) $E^\ominus = +0.15 + 1.28 = +1.43\,V$
(iii) Because the cell was not in standard conditions. **(1)**
2 (a) The fuel in a fuel cell is supplied continuously **(1)**. The voltage from a fuel cell is constant. **(1)**
(b) $2H_2 + O_2 \rightarrow 2H_2O$ **(1)**
$E^\ominus = +0.83 + 0.40 = +1.23\,V$ **(1)**

110. Exam skills 9

1 (a) D
(b)

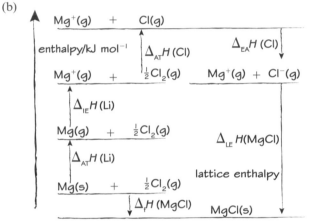

*(cycle: **(3)**)*

$\Delta_f H = 146 + 738 + 122 - 349 - 753$ **(1)**
$= -96\,kJ\,mol^{-1}$ **(1)**

(c) The enthalpy of formation of $MgCl_2$ is more exothermic than that of $MgCl$.
(d) Bubble chlorine at $10^5\,Pa$ **(1)** into a solution of $1\,mol\,dm^{-3}$ chloride ions **(1)** with a platinum electrode **(1)** at $298\,K$ **(1)** connect to SHE and read voltage **(1)**.

111. The transition elements

1 (a) (i) an element with an outer electron in a d orbital
(ii) an element that has an ion with an incomplete d sub-shell
(b) (i) Ti, V, Cr, Mn, Fe, Co, Ni, Cu
(ii) Sc, Zn
(iii) K, Ca, Ga, Ge, As, Se, Br, Kr
2 (a) (i)

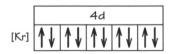

(ii) Cadmium's common ion, Cd^{2+}.

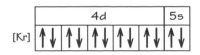

(b) It is not a transition element because its ion has a full d sub-shell.
3 D

112. Properties of transition elements

1 (a) a substance that increases the rate of reaction without getting used up **(1)** by providing an alternative reaction pathway **(1)** with lower activation energy. **(1)**
(b) C
(c) Measure some hydrogen peroxide solution into a flask with a side arm, connect a gas syringe and start the clock, measuring the gas produced in a certain time. **(1)**
Repeat an identical experiment but add a weighed amount of MnO_2. **(1)**
With the MnO_2 the volume of gas produced is more in the same time. **(1)**
Filter after the second experiment and collect and dry the residue, reweighing to show that the catalyst has not been used up. **(1)**
2 (a) the transition element found in A: iron **(1)**
the transition element ion found in solution B: $[Fe(H_2O)_6]^{2+}/Fe^{2+}$ **(1)**
precipitate C: $[Fe(OH)_2(H_2O)_4]/Fe(OH)_2$ **(1)**
substance D: $[Fe(H_2O)_6]^{3+}/Fe^{3+}$ **(1)**
(b) oxidation **(1)**

113. Complex ions

1 (a)

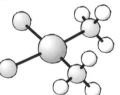

(i) covalent bond **(1)**
(ii) coordinate bond **(1)**
(iii) circle around one NH_3 **(1)**
(b) A
(c) 6 **(1)**, because there are six coordinate bonds. **(1)**
(d) octahedral
2 (a) The nitrogen atoms on the molecule have a lone pair **(1)** that can be donated to the transition metal to form a coordinate bond/ en is bidentate so can form two bonds. **(1)**
(b) A

114. 4-fold coordination and isomerism

1 (a) *cis*

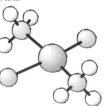

trans

*(Each isomer: **(1)**, isomers correct way round: **(1)**)*
(b) where the complexes have the same structures **(1)** but the ligands are orientated differently in space **(1)**
(c) square planar

(d) (i) e.g. $CuCl_4^{2-}$

(ii)

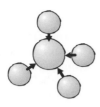

(diagram (1), tetrahedral (1))

2   A

115. Precipitation reactions

1   (a) When two solutions are mixed any solid product is a precipitate.

(b) (i) pale pink/ colourless

(ii) $[Mn(H_2O)_6]^{2+}$

(c) (i) white

(ii) $[Mn(H_2O)_6]^{2+} + 2OH^- \rightarrow [Mn(OH)_2(H_2O)_4] + 2H_2O$/ $Mn^{2+} + 2OH^- \rightarrow Mn(OH)_2$ *(precipitate formula: (1), rest of equation: (1))*

(d) C

(e) no observable change

2   A blue precipitate forms (1), which in excess dissolves (1) to make a deep blue solution. (1)
$[Cu(H_2O)_6]^{2+} + 2OH^- \rightarrow [Cu(OH)_2(H_2O)_4] + 2H_2O$ (1)
$[Cu(OH)_2(H_2O)_4] + 4NH_3 \rightarrow [Cu(NH_3)_4(H_2O)_2]^{2+} + 2H_2O + 2OH^-$/ $Cu(OH)_2 + 4NH_3 + 2H_2O \rightarrow [Cu(NH_3)_4(H_2O)_2]^{2+}$ *(formulae: (1), balancing: (1))*

116. Ligand substitution reactions

1   (a) (i) green precipitate

(ii) $[Cr(H_2O)_6]^{3+} + 3OH^- \rightarrow [Cr(OH)_3(H_2O)_3] + 3H_2O$ *(LHS formulae: (1), RHS formulae: (1), balancing: (1))*

(iii) A

(b) (i) The precipitate dissolves to form a violet solution.

(ii) $[Cr(OH)_3(H_2O)_3] + 6NH_3 \rightarrow [Cr(NH_3)_6]^{3+} + 3OH^- + 3H_2O$ *(LHS formulae: (1)k, RHS formulae: (1), balancing: (1))*

(iii) B

2   (a) Oxygen molecules bind to the molecule of haemoglobin (1) and are carried to the cells where they are released. (1)

(b) CO undergoes a ligand exchange reaction (1) replacing oxygen attached to haemoglobin (1) preventing oxygen transport. (1)

117. Redox reactions of transition elements

1   (a) copper

(b) $[Cu(H_2O)_6]^{2+}$

(c) I_2

(d) CuI

(e) C

2   (a) (i) +6

(ii) potassium dichromate(VI)

(b) from orange (1) to green/ blue/ green-blue (1)

(c) (i) The chromium in dichromate is reduced (1) the alcohol is oxidised (1) the alcohol must be primary or secondary. (1)

118. Exam skills 10

1   (a) (i) Ni [Ar] $3d^8 4s^2$/ [Ar] $4s^2 3d^8$ (1)
Cu [Ar] $3d^{10} 4s^1$/ [Ar] $4s^1 3d^{10}$ (1)
Zn [Ar] $3d^{10} 4s^2$/ [Ar] $4s^2 3d^{10}$ (1)

(ii) Copper ions have an incomplete d sub-shell but zinc ions have a full d sub-shell.

(b) for example:
Acid–base: add sodium hydroxide solution (1) to give a blue precipitate (1)

Ligand exchange: a concentrated hydrochloric acid (1) to give a yellow solution (1)
Redox: add potassium iodide solution (1) to give a white precipitate in a brown liquid (1)

(c) (i)

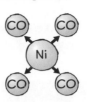

(diagram (1),   diagram (1), tetrahedral (1), square planar (1))

(ii) $[Ni(H_2O)_6]^{2+} + 4CN^- \rightarrow [Ni(CN)_4]^{2-} + 6H_2O$ *(formulae of complex product: (1), rest of equation: (1))*

(iii) Cyanide ions bind to haemoglobin (1) and prevent oxygen joining there (1). When using hydroxocobalamin, cyanide ions bond to this instead. (1)

119. The bonding in benzene rings

1   (a) Molecular formula: C_6H_6 (1) Empirical formula: CH (1)

(b) In a σ bond, the pair of electrons overlap directly between the atoms (1) and in a π bond the overlap is above and below the atoms. (1)

(c) (i) the mixture turns colourless (1)
1,2,3,4,5,6-hexabromocyclohexane (1)
addition reaction (1)
reacts like an alkene with a C=C (1)

(ii) no change

(iii) In benzene the π electrons are delocalised (1) giving a more stable structure so no addition reaction occurs. (1)

(d) (i) $C_6H_6 + 3H_2 \rightarrow C_6H_{12}$

(ii) cyclohexane

120. Reactions of benzene rings

1   (a) C

(b) (i) $FeBr_3 + Br_2 \rightarrow FeBr_4^- + Br^+$ *(LHS: (1), RHS: (1))*

(ii) $C_6H_6 + Br_2 \rightarrow C_6H_5Br + HBr$ *(formulae: (1), balancing: (1))*

(c)

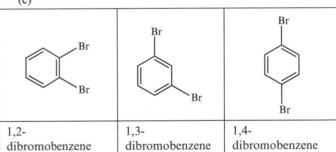

| 1,2-dibromobenzene | 1,3-dibromobenzene | 1,4-dibromobenzene |

(Each structure: (1), three names: (1))

(d) ethanoyl chloride (1), aluminium chloride (1)

121. Electrophilic substitution reactions

1   (a) D

(b) (i) $CH_3CHClCH_3 + AlCl_3 \rightarrow CH_3CH^+CH_3 + AlCl_4^-$

(ii) It is a catalyst (1). It accepts electron density from the Cl / as it is regenerated (1).

(c)

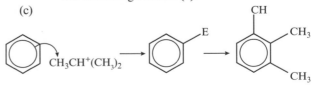

(left hand diagram: (1), carbocation: (1), arrow in middle diagram: (1))

2 2-chloro-2-methylpropane **(1)**

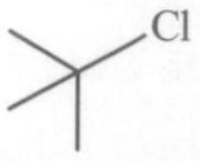

(1)

122. Comparing the reactivity of alkenes and aromatic compounds

1 (a)

hex-2-ene	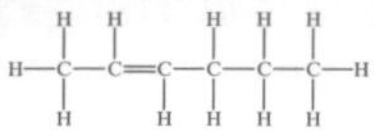 H–C–C=C–C–C–C–H (with H atoms)	bromine decolourised	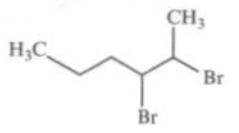 H₃C structure with Br Br
benzene	⬡ (benzene)	no visible change	no product formed
cyclohexene	⬡ (cyclohexene)	bromine decolourised	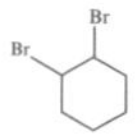 Br Br structure

(All three original structures: (1), each observation: (1), each product: (1))

(b) The electron density in π bonds is delocalised over the ring so is less dense than in hex-2-ene **(1)**, so in benzene no partial charge induced in bromine/ bromine is not a good enough electrophile **(1)**, non-polar bromine does not react with benzene but does with hex-2-ene. **(1)**

2 B

123. Phenol

1 (a) (i) $C_6H_5OH + NaOH \rightarrow C_6H_5O^-Na^+ + H_2O$
(ii)   neutralisation

(b)

(structure of 4-nitrophenol with OH at top and NO₂ at bottom)

(1) 4-nitrophenol **(1)**

(c) (i)   orange bromine solution **(1)** gives a white precipitate **(1)**
(ii)   **B**
(iii)   When bromine reacts with benzene a catalyst (e.g. iron) is required, but no catalyst required with phenol **(1)**. The hydroxyl group releases electrons into the ring **(1)** so phenol is more reactive with electrophiles because of the higher electron density. **(1)**

124. Directing effects in benzene

1 (a) **C**
(b) HBr

2 (a) (i)   concentrated nitric acid **(1)** concentrated sulfuric acid **(1)**
(ii)   acid only has to be dilute **(1)** no sulfuric acid required **(1)**
(iii)   The hydroxyl group releases electrons into the ring **(1)** giving the ring a higher electron density **(1)** so phenol is more reactive with electrophiles. **(1)**
(iv)

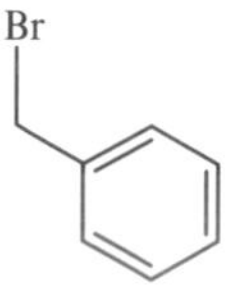

(structure showing Br–CH₂ attached to benzene ring)

1 (a) **B**
(b) (i)   The red/orange of the bromine disappears **(1)** and a white precipitate forms. **(1)**
(ii)   electrophilic substitution **(1)**, 2,4,6-tribromophenylamine **(1)**
(iii)   The lone pair on the N is less available for reaction **(1)** as it forms part of the delocalised system. **(1)**
(iv)

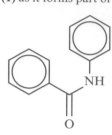

(Amide link: (1), rest of structure: (1))

126. Aldehydes and ketones

1 (a) (i)   A: orange to green **(1)**
B: orange to green **(1)**
C: no visible change **(1)**
D: no visible change **(1)**
(ii)   A: no visible change **(1)**
B: yellow/ orange precipitate **(1)**
C: yellow/ orange precipitate **(1)**
D: no visible change **(1)**
(b) (i)   **A**
(ii) $CH_3CH_2CH(C_2H_5)CH_2CHO + [O]$
$\rightarrow CH_3CH_2CH(C_2H_5)CH_2COOH$ **(1)**
(c)

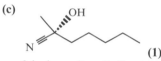

(1)

2-hydroxy-2-methylheptanenitrile **(1)**

127. Nucleophilic addition reactions

1 **C**
2 (a) $NaBH_4$/ $LiAlH_4$
(b) reduction
(c) $CH_3COCH_2CH_3 + 2[H] \rightarrow CH_3CH(OH)CH_2CH_3$
3 (a)

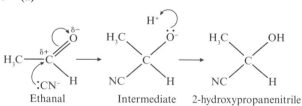

Ethanal   Intermediate   2-hydroxypropanenitrile

(1)

(b) 2-hydroxypropanenitrile **(1)**
(c) The atoms around the carbonyl group are in a plane **(1)**, the cyanide ion attacks with equal probability above and below the plane. **(1)**
(d) HCN(g) is toxic. **(1)**
HCN(aq) is a weak acid giving a low concentration of cyanide ions. **(1)**

128. Carboxylic acids

1 (a) The effervescence is carbon dioxide **(1)** showing that the molecule contains the functional group of carboxylic acid. **(1)**
(b) C = 34.6%, H = 3.85%,
O = 100 − 34.6 − 3.85 = 61.55% **(1)**
amount C = $\frac{34.6}{12}$ = 2.883 mol; H = $\frac{3.85}{1}$ = 3.85 mol **(1)**;
O = $\frac{61.55}{16}$ = 3.847 mol **(1)**
C $\frac{2.883}{2.883}$ = 1; H $\frac{3.85}{2.883}$ = 1.335; O $\frac{3.847}{2.883}$ = 1.334
$C_3H_4O_4$ **(1)**

(c) $M_r(C_3H_4O_4) = 104$, molecular formula = $C_3H_4O_4$ **(1)**

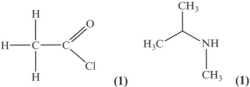

propanedioic acid **(1)**

2   B

3   shake all three with bromine water
the one that forms a colourless mixture is the alkene
add sodium hydrogencarbonate to the remaining two
the one that effervesces is the carboxylic acid
to confirm, add Tollens' reagent to remaining sample
silver mirror forms
*(Each suitable test: **(1)**, result: **(1)**; tests can be in any logical order.)*

129. Esters

1   (a) **D**
(b) non-toxic **(1)** has a distinctive flavour/ smell **(1)**
(c) B
(d) (i) breaking of a bond **(1)** with water **(1)**
(ii) heat **(1)** with dilute acid/ alkali **(1)**
(iii) ethanol   **(1)**

130. Acyl chlorides

1   (a)

(b) $SOCl_2$
(c) (i) $2CH_3CH_2COOH + SOCl_2$
$\rightarrow 2CH_3CH_2COCl + H_2O + SO_2$
(formulae: **1 mark**, balancing: **1 mark**)
(ii) Fume cupboard **(1)** because sulfur dioxide causes breathing problems/ is toxic **(1)**.

2   (a) amide
(b)
(1)   **(1)**
(c) B

131. Amines

1   (a) A: amide
B: amine
C: nitrile
D: nitro
(b) The number of groups joined to the nitrogen is three **(1)** so the molecule is a tertiary amine **(1)**.

2   (a) B
(b) P: $CH_3CH_2CH_2NH_2$
Q: $CH_3CH_2CH_2NHCH_3$
R: HBr
S: $CH_3CH_2CH_2NH_3^+Br^-$

132. Amino acids

1   (a) (i)

(1)

(ii)

(iii) A
(b) A
(c) $CH_2(NH_2)CH_2CH_2CH(NH_2)COOH$
$\rightarrow CH_2(NH_2)CH_2CH_2CH_2NH_2 + CO_2$ **(1)**
1,4-diaminobutane **(1)**

133. Optical isomers

1   (a) (i) carbon dioxide
(ii) carboxylic acid or acyl chloride group is present
(b) (i) oxidation/ redox
(ii) aldehyde **(1)** primary or secondary alcohol **(1)**
(c) (i)

*(correct structure: **(1)**, mirror images: **(1)**; other answers are possible)*
(ii) $CH_2OH)CH_2CH_2COOCl$ **(1)**
(allow any molecule with formula $C_4H_7O_3Cl$ containing either COOH or COCl and a primary or secondary alcohol group)
(d) (i) an equimolar mixture of two optical isomers
(ii) no, because R causes no rotation **(1)** and P and Q isomers' rotations cancel **(1)**.

134. Condensation polymers

1   Both types form a long chain of repeating units from one or two types of monomer. **(1)**
In addition polymerisation no other product is formed but in condensation polymerisation a small molecule is also formed. **(1)**

2   (a) condensation
(b)

diamine	diacyl chloride
(1)	**(1)**
1,4-diaminobenzene **(1)**	

(c) hydrogen chloride/ HCl **(1)**

3   Monomer:

(1)

Repeat unit:

(1)

Type: condensation polymerisation **(1)**

135. Exam skills 12

1 (a) **D**

(b) (i) condensation

(ii) **D**

(c) (i)

Molecule 1: **(1)**

dipotassium benzene-1,4-dicarboxylate **(1)**

Molecule 2: **(1)**

ethane-1,2-diol **(1)**

(molecules can be in either order)

(ii) the dicarboxylic acid forms / not the salt / benzene-1,4-dicarboxylic acid forms

136. Carbon carbon bond formation

1 (a) C = 69.0, N = 16.1, H = 14.9 **(1)**

C: $\frac{69}{12}$ = 5.75, N: $\frac{16.1}{14}$ = 1.15, H: $\frac{14.9}{1}$ = 14.9 **(1)**

C: $\frac{5.75}{1.15}$ = 5, N: $\frac{1.15}{1.15}$ = 1, H: $\frac{14.9}{1.15}$ = 13

C_5NH_{13} **(1)**

(b) $CH_3CH_2C^*H(CH_3)CH_2NH_2$ **(2)**

(c) (i) **D**

(ii) KCN/ NaCN **(1)**

HCl **(1)**

(iii) $CH_3CH_2CH(CH_3)CN + 2H_2$

$\rightarrow CH_3CH_2CH(CH_3)CH_2NH_2$

(formulae: (1), balancing: (1))

(iv) nickel

(v) **B**

137. Purifying organic solids

1 (a) Benzoic acid is more soluble in hot water than in cold water.

(b) Some benzoic acid will also be dissolved in the **cold** water **(1)** so a minimum amount of water must be used to minimise the loss of product / to allow nearly all of the benzoic acid to crystallise out. **(1)**

(c) To remove **soluble** impurities.

(d) The drying is incomplete (step 4) **(1)** so the benzoic acid has an impurity of water. **(1)**

(e) M_r (C_6H_5CHO) = 106; M_r (C_6H_5COOH) = 122 **(1)**

amount benzaldehyde = $\frac{5.00}{106}$ = 0.472 mol **(1)**

maximum mass benzoic acid = 0.472 × 122 = 5.755 g **(1)**

% yield = $\frac{5.35}{6.354}$ × 100 = 93% **(1)**

138. Predicting the properties and reactions of organic compounds

1 (a) alkene **(1)**, ketone **(1)**

(b) (i) mixture turns from orange/yellow to colourless

(ii) no change/ remains orange

(iii) yellow/orange precipitate

(c) **D**

(d) (i) asterisk on bottom carbon of ring

(ii) The different shapes **(1)** bind to different smell receptors. **(1)**

(e)

(**1 mark** for each addition of a water molecule; allow H and OH added to either carbon)

139. Summary of organic reactions

1 (a) A1/ A2: carbon dioxide **(1)** water **(1)**

B: bromine

C: sulfuric acid

D: distil propanal as it forms

E: 1-butylamine

F: Ni

G: concentrated HCl

H: tin

I: pentan-3-one

J: butanoic acid

K: $SOCl_2$ / PCl_5

(b) If the reagents are heated under reflux **(1)** further oxidation (of the aldehyde) occurs **(1)** forming propanone. **(1)**

(c) 1, 2, 3, 4, 5, 6

(5 or 6 of these without 7 = (2), 4 of these without 7 = (1))

140. Organic synthesis

1 (a) B, A, D, C

(b) **Step 1:** from benzene (**B**) to chlorobenzene (**A**) **(1)**

chlorine **(1)**

$AlCl_3$ / halogen carrier **(1)**

Step 2: from chlorobenzene (**A**) to 4-chloronitrobenzene (**D**) **(1)**

concentrated nitric acid **(1)**

concentrated sulfuric acid **(1)**

Step 3: from 4-chloronitrobenzene (**D**) to 4-chlorophenylamine (**C**) **(1)**

concentrated hydrochloric acid **(1)**

tin **(1)**

141. Thin layer chromatography

1 (a) **D**

(b) (i) As time passes, the spot showing R disappears and the spot showing P goes from faint to strong **(1)** showing R changing to P **(1)**.

(ii) to compare the height of the spots so that R and P can be identified **(1)**

(iii) distances: R = 27 mm; P = 20 mm; solvent front = 31 mm **(1)**, R_f(R) = 27/31 = 0.87 **(1)**, R_f (P) = 20/31 = 0.65 **(1)**. (3)

2 **D**

142. Gas chromatography

1 (a) There are three components. **(1)**

A has lowest retention time and C the highest. **(1)**

A has least/ C has most strong interaction with stationary phase. **(1)**

A is in the smallest amount and B in the largest. **(1)**

2 (a) (i) helium/ nitrogen/ other unreactive gas

(ii) stationary phase

(b) **C**

(c) mass spectrometer

143. Qualitative tests for functional groups (1)

1. (a) add sodium carbonate/ sodium hydrogencarbonate/ a named indicator **(1)**
   effervescence with **A** / correct colour of indicator **(1)**
   (b) (i) add 2,4-dinitrophenylhydrazine **(1)**
      yellow/orange precipitate only with **C** and **D** **(1)**
   (ii) warm with Tollens' reagent **(1)**
      silver mirror only with **C** **(1)**
   (c) warm with acidified potassium dichromate **(1)**
      B goes green **(1)**
      E stays orange **(1)**
   (d) **C**

144. Qualitative tests for functional groups (2)

1. (a) (i) CO_2
      (ii) **B**
   (b) AgBr **(1)**
      cream **(1)**
   (c) (i) nucleophilic substututation/ hydrolysis
      (ii) orange to green
      (iii) alcohol **(1)**
      primary or secondary **(1)**
   (d)

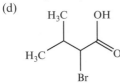

   (correct molecule: (1); molecule with COOH and Br and correct molecular formula: (1))

145. Carbon-13 NMR spectroscopy

1. (a) **B**
   (b)

$$CH_3CH_2 - \overset{\overset{\textstyle O}{\|}}{C} - \overset{\overset{\textstyle CH_3}{|}}{\underset{\underset{\textstyle CH_3}{|}}{\underset{O-CH}{}}}$$

   (ester with formula $C_6H_{12}O_2$: (1), fully correct : (2): allow above and ethyl 2-methylpropanoate and methyl 3-methyl butanoate)
   (c)

   OH
   1
   2   2
   3   3
   4

   (structure: (1))
   The number of different carbon environments is four **(1)** and these are labelled on the diagram 1, 2, 3, 4 **(1 for numbers on diagram) (1)**
   likely chemical shift 110–160 ppm **(1)**.

146. Proton NMR spectroscopy

1. (a) propan-1-ol: 4 peaks **(1)**
   propan-2-ol: 3 peaks **(1)**
   spectrum is of propan-2-ol **(1)**
   (b) Peak **A** is split into 7 so has 6 neighbours **(1)** so is the CH hydrogen **(1)**
   Peak **B** is unsplit, a singlet **(1)** and due to the OH hydrogen **(1)**
   Peak **C** is a doublet **(1)** so is due to methyl group hydrogens **(1)**
   Ratio **A** : **B** : **C** is 1 : 1 : 6 **(1)**

---

2. (a)

   H   H
      C
   H—C   C—H
   H—C   C—H
      C
   H   H

   (b) (i) 1 peak, singlet
      (ii) 3 peaks **(1)**
         1 : 2 : 2 **(1)**

147. Identifying the structure of a compound from a proton (H-1) NMR spectrum

1. (a) **D**
   (b) The left-hand peak, the CH_2 protons, is a quartet as it is split by three CH_3 protons. **(1)**
   The middle peak, the CH_3 protons, is a triplet as it is split by two CH_2 protons. **(1)**
   (c) (i) There are three peaks so there are hydrogens in three environments **(1)** so it is 1-bromopropane. **(1)**
      (ii) the CH_2 protons on middle carbon
      (iii) the integration trace / the chemical shift **(1)**
         as the CH_3 protons to CH_2Br protons would have ratio 3 : 2 / as the CH_3 protons and CH_2Br protons would have different shift values **(1)**

148. Predicting a proton NMR spectrum

1. (a) **S**
   (b) **B**
   (c) **D**
   (d) **R**
2. The CH_3 and CH_2 hydrogens produce identical peaks **(1)**
   the OH proton becomes OD **(1)** so the peak on the far left disappears. **(1)**

149. Deducing the structure of a compound from a range of data

1. (a) (i) C 62.1; O 27.6; H 10.3 **(1)**
      $C = \frac{62.1}{12} = 5.175$; $O = \frac{27.6}{16} = 1.725$; $H = \frac{10.3}{1} = 10.3$ **(1)**
      $C = \frac{5.175}{1.725} = 3$; $O = \frac{1.725}{1.725} = 1$; $H = \frac{10.3}{1.725} = 5.97$: C_3OH_6 **(1)**
      (ii) the molecular ion shows that $M_r = 116$, $C_3OH_6 = 58$ **(1)** molecular formula = $C_6O_2H_{12}$ **(1)**
   (b) has C=O and O–H and only one functional group **(1)**, so carboxylic acid **(1)**
   (c)

   O
   ‖
   HO—C—
   (1)
   2,2-dimethylbutanoic acid **(1)**
   (d) A: CH_2 protons
   B: CH_3 protons for methyl at end of chain
   C: OH proton
   D: protons on two methyl substituents

150. Exam skills 13

1. (a) **T** is a primary alcohol **(1)**, secondary alcohol **(1)** or aldehyde **(1)**
   (allow (1) for alcohol without specifying classification)
   (b) **T** is not an aldehyde/ it is a primary or secondary alcohol
   (c) A carboxylic acid would cause effervescence **(1)**.
   A primary alcohol/ aldehyde would be oxidised to a carboxylic acid **(1)**, so **T** is not a primary alcohol (or aldehyde). **(1)**

(d) **T** is a secondary alcohol.
(e)

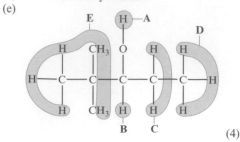

(4)

(−1 for each error)

AS Practice paper 1

SECTION A
(**1 mark each**)

1	D	2	C	3	B	4	C	5	D	6	B
7	B	8	B	9	B	10	D	11	B	12	C
13	C	14	A	15	C	16	C	17	D		
18	B	19	A	20	D						

SECTION B

1 (a) $1s^2 2s^2 2p^6 3s^2 3p^5$

(b)

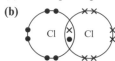

(Shared pair: (1), rest of molecule: (1))

(c) (i) atoms with the same number of protons. same atomic number (1) but different numbers of neutrons/ different mass number (1)

(ii) $(35 \times \frac{75}{100}) + (37 \times \frac{25}{100})$ (1)
= 35.5 (1)

(iii) high energy electrons (1)
knock electron out of outer shell of molecule (1)
Cl_2^+ / Cl^+ (1)

(d) make a solution with the solid (1)
add nitric acid (1)
add silver nitrate solution (1)
a white precipitate forms (1)

(e) (i) Br_2 (1)

(ii) $Cl_2 + 2I^- \rightarrow 2Cl^- + I_2$ *(formulae: (1), balancing: (1))*

(iii) no observation (1)
because no reaction occurs as chlorine is less reactive than fluorine (1)

2 (a) $3C(s) + 4H_2(g) \rightarrow C_3H_8(g)$ *(formulae: (1), state symbols: (1))*

(b) enthalpy of combustion of reactants = $(3 \times -394 + 4 \times -286)$ (1)
$\Delta_c H^\ominus = -2326 + 2220$ (1)
$= -106 \, kJ \, mol^{-1}$ (1)

(c) combustion is incomplete

3 (a) Any two from *(structures: (1) each, associated name: (1) each)*

Cl H H | | | Cl—C—C—C—H | | | H H H	Cl Cl H | | | H—C—C—C—H | | | H H H
1,1-dichloropropane	1,2-dichloropropane
Cl H Cl | | | H—C—C—C—H | | | H H H	H Cl H | | | H—C—C—C—H | | | H Cl H
1,3-dichloropropane	2,2-dichloropropane

(b) (i) $CH_3CHBrCH_2CH_3 + OH^- \rightarrow CH_3CH(OH)CH_2CH_3 + Br^-$ *(alcohol product: (1), rest of equation: (1))*

(ii) hydroxide ion/ OH^-

(iii) alcohol

(c) haloalkanes release halogen radicals (1)
which catalyse ozone decomposition (1)
which allows UV light to reach Earth (1)

4 (a) oppositely charged ions (1)
arranged in a regular way (1)

(b) ions can move only when freed from lattice

(c) (i) anhydrous copper(II) sulfate

(ii) to ensure that all water has been removed

(iii) mass water = 2.50 − 1.60 = 0.90 g (1)
$M_r(CuSO_4) = 159.6$ (1)
amount $CuSO_4 = \frac{1.60}{159.6} = 0.0100 \, mol$; amount water
$= \frac{0.90}{18} = 0.05 \, mol$ (1)
ratio $= \frac{0.05}{0.01} = 5$, so $x = 5$ (1)

5 (a) $CaCO_3 + 2HCl \rightarrow CaCl_2 + H_2O + CO_2$ *(formulae: (1), balancing: (1))*

(b) $Mr \, (CaCl_2) = 111$ (1)
total of products = 111 + 18 + 44 = 173 (1)
atom economy
$= \frac{111}{173} \times 100\% = 64\%$ (1)

AS Practice paper 2

1 (a) $MCO_3 \rightarrow MO + CO_2$

(b) to ensure that the reaction is complete

(c) Mass = 12.729 − 11.782 = 0.947 g (1)
amount of carbon dioxide = 0.947/ 44 = 0.021 523 mol (1)

(d) (i) 11.782 − 9.552 = 2.23 g (1)

(ii) $M_r = \frac{2.23}{0.021 \, 523}$ (1)
= 103.6 (1)
MO is 103.56, so $A_r(\mathbf{M})$ is 103.6 − 16 = 87.6 (1), **M** is strontium (1)

(e) Some oxide is lost from the top of the crucible.

2 (a) almost all of the molecules dissociate to give H^+ ions

(b) $H^+(aq) + OH^-(aq) \rightarrow H_2O(l)$ *(formulae: (1), state symbols: (1))*

(c) (i) $q = mc\Delta T = 50 \times 4.18 \times 13.1$ (1)
= 2737.9 = 2.74 kJ (1)

(ii) $\frac{25}{1000} \times 2.0$ (1)
= 0.050 mol (1)

(d) (i) enthalpy change when one mole of water is formed in a neutralisation reaction

(ii) $\frac{-2.74}{0.050}$ (1)
$= -54.8 \, kJ \, mol^{-1}$ (1)

(e) use a polystyrene cup (1)
add a lid (1)

3 (a) The silicon is in a giant lattice (1)
where atoms are joined by strong covalent bonds (1)
that require a lot of energy to break. (1)

(b) Sulfur and phosphorus consist of molecules where only the intermolecular forces need to be overcome to melt the substance (1), and as sulfur has more electrons in its S_8 molecule than phosphorus in its P_4 molecule, sulfur has stronger intermolecular forces. (1)

(c) These metals consist of lattices of cations (1) with delocalised electrons that are free to move when a potential is applied. (1)

(d) (i) The particles gain energy **(1)** so that a higher proportion of them have at least the activation energy **(1)** so that a greater proportion of the collisions are successful. **(1)**

(ii)

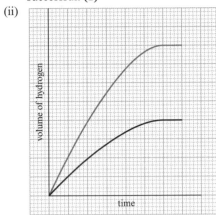

(iii) Hydrogen has a very low density/ mass of hydrogen low.

4 (a)

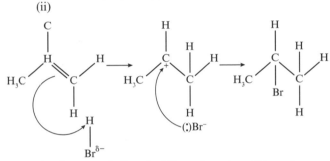

(b) (i) overlap of atomic orbitals **(1)**
directly between the atoms **(1)**

(ii)

(c) (i) 1-bromopropane **(1)**
2-bromopropane **(1)**

(ii)

*(curly arrows in Step 1: **(1)**; carbocation: **(1)**, curly arrows in step 2: **(1)**)*
mechanism: electrophilic **(1)** addition **(1)**

5 (a) concentrated sulfuric acid

(b) from orange **(1)**
to green **(1)**

(c)

Structure	Primary, secondary or tertiary	Oxidation product
OH	tertiary **(1)**	no reaction **(1)**
OH	primary **(1)**	butanoic acid **(1)**
OH	secondary **(1)**	butanone **(1)**

(d) (i) hydrogen bonding

(ii) O–H bonds are polar due to electronegativity difference **(1)** but C–H bonds are not polar (enough) **(1)**

6 (a) (i) $4HCl + O_2 \rightarrow 2Cl_2 + 2H_2O$ *(formulae: **(1)**, balancing: **(1)**)*

(ii) amount HCl $= \dfrac{1000}{36.5} = 27.397\,\text{mol}$ **(1)**

expected amount chlorine $= \dfrac{27.397}{2} = 13.6986\,\text{mol}$ **(1)**

amount chlorine $= \dfrac{775}{71} = 10.915\,\text{mol}$

yield $= \dfrac{10.915}{13.6985} \times 100\% = 80\%$ **(1)**

(iii) $V = \dfrac{nRT}{P}$ **(1)**

$n = \dfrac{775}{71} = 10.915\,\text{mol}$ **(1)**

$T = 400 + 273 = 673\,\text{K}$ **(1)**

$V = 10.915 \times 8.31 \times \dfrac{673}{10^5} = 0.610\,\text{m}^3$ **(1)**

(b) $1s^2 2s^2 2p^6 3s^2 3p^5$

(c) (i) The electronegativity of the two atoms is different.

(ii) The polar bonds are symmetrically arranged so the polarity cancels.

A Level Practice paper 1

SECTION A

(1 mark each)

1	C	2	C	3	D	4	B	5	B	6	B
7	D	8	D	9	D	10	A	11	C		
12	B	13	B	14	D	15	B				

SECTION B

1 (a) $1s^2 2s^2 2p^6 3s^2 3p^5$

(b) (i) Br_2 **(1)**

(ii) $Cl_2 + 2I^- \rightarrow 2Cl^- + I_2$ *(formulae: **(1)**, balancing: **(1)**)*

(iii) no observation **(1)**
because no reaction occurs as chlorine is less reactive than fluorine **(1)**

2 (a) $3C(s) + 4H_2(g) \rightarrow C_3H_8(g)$ (formulae: **(1)**, state symbols: **(1)**)

(b) enthalpy of combustion of reactants $= (3 \times -394 + 4 \times -286)$ **(1)**
$\Delta_c H^{\ominus} = -2326 + 2220$ **(1)**
$-106\,\text{kJ mol}^{-1}$ **(1)**

(c) combustion is incomplete **(1)**

3 (a) $MCO_3 \rightarrow MO + CO_2$ **(1)**

(b) To ensure that the reaction is complete. **(1)**

(c) Mass $= 12.729 - 11.782 = 0.947\,\text{g}$ **(1)**

amount of carbon dioxide $= \dfrac{0.947}{44} = 0.021\,523\,\text{mol}$ **(1)**

(d) (i) $12.729 - 10.500 = 2.229\,\text{g}$ **(1)**

(ii) $M_r = \dfrac{2.229}{0.021523}$ **(1)**
$= 103.56$ **(1)**
MO is 103.56, so M is $103.56 - 16 = 87.56$, M is strontium **(1)**

(e) Some oxide is lost from the top of the crucible. (1)

4 (a) Almost all of the molecules dissociate to give H^+ ions. (1)

(b) $H^+(aq) + OH^-(aq) \rightarrow H_2O(l)$ (formulae: **1 mark**, state symbols: **1 mark**)

(c) (i) $q = mc\Delta T = 50 \times 4.18 \times 13.1$ **(1)**
$= 2737.9 = 2.74\,\text{kJ}$ **(1)**

(ii) $\dfrac{25}{1000} \times 2.0$ **(1)**
$= 0.050\,\text{mol}$ **(1)**

(d) (i) Enthalpy change when one mole of water is formed in a neutralisation reaction. **(1)**

(ii) $\dfrac{-2.74}{0.050}$ **(1)**

$= -54.8\,\text{kJ mol}^{-1}$ **(1)**

(e) use a polystyrene cup **(1)**

add a lid **(1)**

5 (a) The metal consists of a lattice of positive ions/ cations **(1)** with delocalised electrons that are free to move when a potential is applied. **(1)**

(b) (i) The particles gain energy **(1)**

so that a higher proportion of them have at least the activation energy **(1)**

so that a greater proportion of the collisions are successful. **(1)**

(ii)

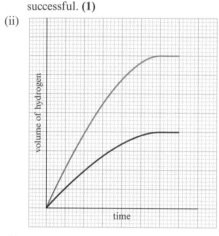

(2)

(iii) Hydrogen has a very low density. **(1)**

6 (a) The iron ions gain electrons. **(1)**

(b) hydrogen gas at 10^5 Pa/ 1 bar **(1)**

is bubbled into a solution containing $1\,\text{mol dm}^{-3}$ H^+ ions **(1)**

with a platinum electrode at 298 K **(1)**

$E^{\ominus} = 0.00\,\text{V}$ **(1)**

(c) chlorine **(1)**

the reduction of chlorine has the most positive $E^{\ominus}$ value **(1)**

it is the smallest atom **(1)**

so attracts an added electron the most **(1)**

(d) VO^{2+} +4 **(1)**

V^{3+} +3 V^{2+} +2 **(1)**

(e) (i) $Fe(s) + 2VO^{2+}(aq) + 4H^+(aq) \rightarrow Fe^{2+}(aq) + 2V^{3+}(aq) + 2H_2O(l)$ (formulae: **(1)**, balancing: **(1)**)

$Fe(s) + 2V^{3+} \rightarrow Fe^{2+}(aq) + 2V^{2+}(aq)$ (formulae: **(1)**, balancing: **(1)**)

(ii) first reaction: $+0.44 +0.34 = 0.78\,\text{V}$ **(1)**

second reaction: $+0.44 -0.26 = 0.18\,\text{V}$ **(1)**

7 (a) (i) a blue precipitate **(1)**

which dissolves **(1)**

to form a deep blue solution **(1)**

(ii) ligand exchange **(1)**

(b) (i) $K_{\text{stab}} = \dfrac{[[Cu(NH_3)_2(H_2O)_4]^{2+}]}{[[Cu(H_2O)_6]^{2+}][NH_3]^2}$ **(1)**

(ii) $\text{dm}^6\,\text{mol}^{-2}$ **(1)**

(c) (i) In each case similar bonds are broken and made **(1)** which are two Cu–O coordinate bonds broken and two Cu–N coordinate bonds made. **(1)**

(ii) In the first reaction there is no change in the number of particles but in the second there is an increase (**(1)** giving an increase in entropy making the forward reaction more likely. **(1)**

8 (a) a high pressure **(1)**

because there are two gas moles on the right and three on the left so the equilibrium position moves to the right with a higher pressure **(1)**

a low temperature **(1)**

because the forward reaction is exothermic so the equilibrium position moves to the right with a lower temperature **(1)**

(b) (i) $K_C = \dfrac{[NO_2]^2}{[NO]^2[O_2]}$ **(1)**

(ii) at equilibrium, amount NO = 0.3 mol **(1)**

amount NO_2 = 0.9 mol, amount $O_2 = 0.8 - \dfrac{0.9}{2} = 0.35\,\text{mol}$ **(1)**

$K_c = \dfrac{\left(\dfrac{0.9}{2}\right)^2}{\left(\dfrac{0.3}{2}\right)^2\left(\dfrac{0.35}{2}\right)}$ **(1)**

$= 51\,\text{dm}^3\,\text{mol}^{-1}$ **(1)**

(c) (i) NO second order **(1)**

O_2 first order **(1)**

(ii) rate = $k[NO]^2[O_2]$ **(1)**

(iii) $k = \dfrac{2.1 \times 10^{-7}}{(1.5 \times 10^{-5})^2(0.5 \times 10^{-5})}$ **(1)**

$= 1.9 \times 10^8$ **(1)**

units $= \dfrac{(\text{mol dm}^{-3}\,\text{s}^{-1})}{(\text{mol dm}^{-3})^2(\text{mol dm}^{-3})} = \text{dm}^6\,\text{mol}^{-2}\,\text{s}^{-1}$ **(1)**

(iv) Step 2 **(1)** because step 1 has no oxygen, but oxygen is in the rate equation **(1)**

A Level Practice paper 2

SECTION A

(1 mark each)

1	B	2	C	3	C	4	C	5	B	6	A
7	D	8	B	9	A	10	B	11	D	12	D
13	C	14	D	15	C								

SECTION B

1 (a) Any two from (structures: **(1)** each, associated name: **(1)** each)

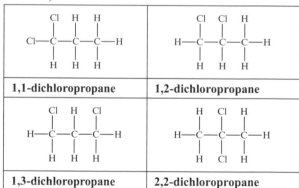

| 1,1-dichloropropane | 1,2-dichloropropane |
| 1,3-dichloropropane | 2,2-dichloropropane |

(b) (i) $CH_3CHBrCH_2CH_3 + OH^- \rightarrow CH_3CH(OH)CH_2CH_3 + Br^-$ (alcohol product: **(1)**, rest of equation: **(1)**)

(ii) hydroxide ion

(iii) alcohol

(c) haloalkanes release halogen radicals **(1)**

which catalyse ozone decomposition **(1)**

which allows UV light to reach Earth **(1)**

2 (a) (i) $HCOOH + CH_3CH_2CH_2OH \rightleftharpoons HCOOCH_2CH_2CH_3 + H_2O$ (LHS formulae: **(1)**, RHS formulae: **(1)**)

(ii) concentrated sulfuric acid

(iii)

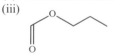

(structure: **(1)**)

propyl methanoate **(1)**

(b) M_r (ester) = 88 **(1)**

total of products = 88 + 18 = 106 **(1)**

atom economy $= \dfrac{88}{106} \times 100\% = 83\%$ **(1)**

(c) (i) trigonal planar **(1)**
 any angle >120° but less than 130° **(1)**
 (ii) C_2H_4O
(d) (i) ethanoic acid **(1)**
 ethanol **(1)**
 ethanoic anhydride **(1)**
 (ii) higher yield **(1)**
 because not an equilibrium reaction **(1)**
 (allow reaction is faster for **(1)**)
 (iii) reactants must be dry **(1)**
 because the anhydride reacts with water **(1)**

3 (a) The acid is in excess. **(1)**
(b) Wear gloves as the phosphoric acid is corrosive. **(1)**
(c)

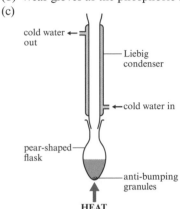

 (flask and heat: **(1)**, water in at bottom and out at top: **(1)**, apparatus not sealed: **(1)**)
(d) the reaction of the unreacted ethanoic anhydride with water **(1)**
 this is to remove most of the acid anhydride before the mixture is poured into the beaker **(1)**
(e)

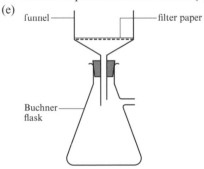

 (correct funnel and filter paper: **(1)**, side-arm flask: **(1)**)

(f) (i) It can form H bonds with water.
 (ii) because aspiring is very soluble in hot water **(1)**
 but less soluble in cold water **(1)**
 (iii) so less aspirin is left in cold solution / so that most of the aspirin crystallises out **(1)**
 to maximise yield **(1)**
(g) (i) carboxylic acid **(1)**
 alcohol **(1)**
 ester **(1)**
 (ii) amount 2-hydroxybenzoic acid $= \dfrac{(63.28 - 58.35)}{138} = 0.0357\,\text{mol}$ **(1)**

 amount (aspirin) $= \dfrac{4.37}{180} = 0.0243\,\text{mol}$ **(1)**

 % yield $= \dfrac{0.0243}{0.0357} \times 100\%$ **(1)**

 $= 68\%$ **(1)**
 (iii) aspirin is lost in transfers/ purification/ remains dissolved in the cold water used in recrystallisation (1)
(h) 3200–3600 cm⁻¹ **(1)**
 the phenolic O–H peak **(1)**

4 (a) (i) $C = \dfrac{55.2}{12}$ $O = \dfrac{18.4}{16}$ $N = \dfrac{16.1}{14}$ $H = \dfrac{10.3}{1}$ **(1)**
 C = 4.6 O = 1.15 N = 1.15 H = 10.3 **(1)**
 C_4ONH_9 **(1)**
 (ii) C_4ONH_9

(b) (i)
 (amine group can be on carbon 2 or 3) **(1)**

 (ii)
 or
 (1)

 (iii)
 or

(c) add Tollens' reagent and warm / add UI **(1)**
 molecule in (ii) gives silver precipitate / UI blue **(1)**
 molecule in (iii) gives no precipitate / UI green **(1)**

5 (a) A: benzene
 B: nitrobenzene
 C: phenylamine (allow aniline)
 D: phenylammonium chloride
(b) (i) concentrated nitric acid **(1)**
 concentrated sulfuric acid **(1)**
 (ii) electrophilic substitution **(1)**
 (iii) $HNO_3 + 2H_2SO_4 \rightarrow NO_2^+ + H_3O^+ + 2HSO_4^-$ **(1)**

 (curly arrow to NO_2^+: **(1)**, + charge and partial ring: **(1)**, curly arrow from C–H to ring: **(1)**)
(c) (i) tin **(1)**
 concentrated hydrochloric acid **(1)**
 (ii) reduction
(d) (i) 78
 (ii) 77
 (iii) NO_2^+

A Level Practice paper 3

1 (a) (i) $CaO(s) + H_2O(l) \rightarrow Ca(OH)_2(aq)$ (formulae: **(1)**, state symbols: **(1)**)
 (ii) the white solid reduces in volume/ disappears **(1)**
 some steam is evolved **(1)**
 (iii) to neutralise acidic soils
(b) $Ca(OH)_2(aq) + CO_2(g) \rightarrow CaCO_3(s) + H_2O(l)$ (formulae: **(1)**, state symbols: **(1)**, $CaCO_3$ underlined: **(1)**)
(c) (i) some impurities may be toxic **(1)**
 so sample cannot be in uses such food/ medicines **(1)**
 (ii) amount HCl $= \dfrac{48.9}{1000} \times 1.50 = 0.07335\,\text{mol}$ **(1)**

 amount $CaCO_3 = \dfrac{0.07335}{2} = 0.036\,675\,\text{mol}$ **(1)**

 mass $CaCO_3 = 100 \times 0.036\,675 = 3.6675\,\text{g}$ **(1)**

 percentage $= \dfrac{3.6675}{3.76} \times 100 = 97.5\%$ **(1)**
 (iii) They do not react with the acid.
(d) (i) $K_p = p(CO_2)$
 (ii) $\Delta S^\ominus = (40.0 + 214.0) - (92.9) = 161.1\,\text{J K}^{-1}\text{mol}^{-1}$ **(1)**
 $\Delta G = \Delta H - T\Delta S = 0$ **(1)**
 $T = \dfrac{\Delta H}{\Delta S} = \dfrac{179}{0.1611}$ **(1)**
 $= 1111\,\text{K}$ **(1)**

2 (a) acid is pipetted into a flask **(1)**
alkali is added from a burette in known amounts **(1)**
the pH is measured with a pH probe **(1)**

(b) acid weak, alkali strong

(c) original amount ethanoic acid $= \frac{100}{1000} \times 0.120 = 0.012\,mol$

amount potassium hydroxide $= \frac{20}{1000} \times 0.100 = 0.002\,mol$ **(1)**

after neutralisation, [ethanoic acid] $= \frac{(0.012 - 0.002)}{0.120} = 0.083\,33\,mol\,dm^{-3}$ **(1)**

[potassium ethanoate] $= \frac{0.002}{0.120} = 0.016\,67\,mol\,dm^{-3}$ **(1)**

$[H^+] = 1.76 \times 10^{-5} \times \frac{0.083333}{0.01667} = 8.8 \times 10^{-5}\,mol\,dm^{-3}$ **(1)**

pH $= -\log(8.8 \times 10^{-5}) = 4.06$ **(1)**

3 (a) (i) +4 **(1)**

(ii) −1 **(1)**

(b) by providing an alternative reaction pathway **(1)**
of lower activation energy **(1)**

(c) (i) [Ar] $3d^5$

(ii) pink/ colourless

4 (a)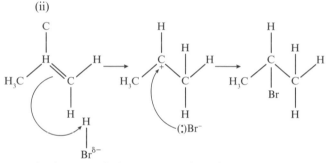

(b) (i) overlap of atomic orbitals **(1)**
directly between the atoms **(1)**

(ii)

(c) 1-bromopropane **(1)**
2-bromopropane **(1)**

(ii)

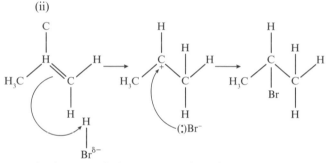

(curly arrows in Step 1: **(1)**; carbocation: **(1)**, curly arrows in Step 2: **(1)**)
mechanism: electrophilic **(1)** addition **(1)**

5 (a) (i) $4HCl + O_2 \rightarrow 2Cl_2 + 2H_2O$ (formulae: **(1)**, balancing: **(1)**)

(ii) amount HCl $= \frac{1000}{36.5} = 27.397\,mol$ **(1)**

expected amount chlorine $= \frac{427.397}{2} = 13.6985\,mol$ **(1)**

amount chlorine $= \frac{775}{71} = 10.915\,mol$

yield $= \frac{10.915}{13.6985} = 80\%$ **(1)**

(b) The polar bonds are symmetrically arranged so the polarity cancels / the molecule is symmetrical so the polarity cancels.

6 (a) A CH_3CH_2COOH
B CO_2
C SO_2
D CH_3CH_2COCl
E $CH_3CH_2COOCH_3$
F HCl

(b) (i) $CH_2OHCH=CHCH_2OH$ (any four carbon alkene with two OH groups)

(ii) C=C

(c) (i) $HOOCCH_2CH_2CH(NH_3^+Cl^-)COOH$

(ii) $NaOOCCH_2CH_2CH(NH_2)COO^-Na^+$

(iii)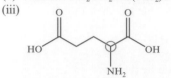

(d) (i) It is only present in small amounts in tomatoes.

(ii) heat to remove water and increase concentration of the glutamate

(iii) dissolve solid in minimum quantity hot solvent **(1)**
filter hot and crystallise **(1)**
refilter, wash and dry **(1)**

(iv) find melting point **(1)**
which should be sharp and match data book value **(1)**